Other books of interest

The ICE Conditions: Sixth Edition
A User's Guide
Brian Eggleston
0–632–03092–5

The ICE Design and Construct Contract
A Commentary
Brian Eggleston
0–632–03697–4

Liquidated Damages and Extensions of Time
In Construction Contracts
Brian Eggleston
0–632–03295–2

Design Liability in the Construction Industry
Fourth Edition
D.L. Cornes
0–632–03261–8

In preparation

Claims under the ICE Conditions
Brian Eggleston
0–632–03912–4

Building Sub-Contracts
A Guide for Specialist Contractors
S.T.R. Klein
0–632–03763–6

Model Forms of Contract for Electrical and Mechanical Plant

Brian Eggleston
CEng, FICE, FIStructE, FCIArb

**Blackwell
Science**

© Brian Eggleston 1995

Blackwell Science Ltd
Editorial Offices:
Osney Mead, Oxford OX2 0EL
25 John Street, London WC1N 2BL
23 Ainslie Place, Edinburgh EH3 6AJ
238 Main Street, Cambridge
  Massachusetts 02142, USA
54 University Street, Carlton
  Victoria 3053, Australia

Other Editorial Offices:
Arnette Blackwell SA
1, rue de Lille, 75007 Paris
France

Blackwell Wissenschafts-Verlag GmbH
Kurfürstendamm 57
10707 Berlin, Germany

Blackwell MZV
Feldgasse 13, A-1238 Wien
Austria

First published 1995

Set by DP Photosetting, Aylesbury, Bucks
Printed and bound in Great Britain by
Hartnolls Ltd., Bodmin, Cornwall

DISTRIBUTORS

Marston Book Services Ltd
PO Box 87
Oxford OX2 0DT
(*Orders:* Tel: 01865 791155
        Fax: 01865 791927
        Telex: 837515)

USA
Blackwell Science, Inc.
238 Main Street
Cambridge, MA 02142
(*Orders:* Tel: 800 215-1000
            617 876-7000
        Fax: 617 492-5263)

Canada
Oxford University Press
70 Wynford Drive
Don Mills
Ontario M3C 1J9
(*Orders:* Tel: 416 441 2941)

Australia
Blackwell Science Pty Ltd
54 University Street
Carlton, Victoria 3053
(*Orders:* Tel: 03 347-5552)

A catalogue record for this title
is available from the British Library

ISBN 0–632–03803–9

Library of Congress
Cataloging-in-Publication Data

Eggleston, Brian.
    Model forms of contract for electrical
  and mechanical plant/Brian Eggleston
        p.   cm.
    Includes index.
    ISBN 0–632–03803–9
    1. Electric engineering contracts—Great
Britain.   2. Mechanical engineering
contracts—Great Britain.   I. Title.
KD1641.E35   1995
343.73'0786213—dc20
[347.303786213]   94-46514
                                      CIP

Contents

Preface

Some years ago when my consultancy work in the water industry first brought me into regular contact with the model forms for plant and process contracts I became aware how little there was by way of guidance in published works of reference. And I soon realized that many users of the model forms, myself included, were attempting to operate, interpret and in some cases amend the forms without full appreciation of their special characteristics and their underlying philosophy on the risks, responsibilities and liabilities of the parties.

But I also found widespread and keen interest to learn about the model forms stimulated greatly by the initiative of the Joint IMechE/IEE Committee and the Water Services Association in producing a new range of model forms for plant contracts. As a result I have come to spend nearly as much time giving advice and training courses on plant and process contracts as I do on construction contracts and I am regularly asked which books I recommend. I wish there was a list but there is not.

My objective therefore in writing this book is to ensure that there is at least one commentary available on the model forms for plant contracts. I obviously hope that in a modest way this book will contribute to better understanding of the forms so I have gone some way beyond straight analysis of text and tried to show how the model forms compare with other standard forms of contract and how they fit in with general legal principles.

I am indebted to Mr Dan Graham, Partner in Messrs Bristows Cooke & Carpmael, Solicitors, and former Chairman of the Drafting Panel for the IMechE/IEE model forms, for his time in kindly explaining to me the intentions behind various provisions in the forms whenever I ran into difficulties.

Brian Eggleston
5 Park View
Arrow
Alcester
Warwicks B49 5PN

Acknowledgements

The extracts from the texts of Model Forms, MF/1, MF/2 and MF/3 are reproduced by kind permission of the copyright holders – The Institution of Electrical Engineers.

The extracts from the text of the General Conditions of Contract for Water Industry Plant Contracts, Form G/90, are reproduced by kind permission of the Water Services Association.

Author's note

Phraseology

Rather than repeat throughout the book the full titles of the Model Forms I have resorted to abbreviations. Generally I refer to MF/1, MF/2, MF/3 and G/90.

Capitals

As a matter of style capital letters have been used sparingly. The result, I hope, is easier reading. However, it does mean that defined contractual terms and titles such as the purchaser, the engineer and the contractor appear completely in lower case.

Text of the Model Forms

For commercial reasons it has not been possible to include the full text of each of the Model Forms in this book. I have assumed, therefore, that readers will have to hand a copy of the appropriate form for any particular point of reference.

The text which is quoted is taken from the following editions of the model forms:

- MF/1 – 1992 reprint of 1988 edition
- MF/2 – 1991 edition
- MF/3 – 1993 edition
- G/90 – October 1990 edition

Grouping of the Model Forms for Commentary

Readers will see from the contents schedule of this book that in commenting on the various model forms I have selected different groupings of the model forms in some chapters than in others. This is because the similarities between the forms vary from topic to topic and consistent grouping would have unnecessarily extended the text.

Chapter 1

Introduction

1.1 Model forms for plant contracts

The model forms for electrical and mechanical plant examined in detail in this book are:

- MF/1
- MF/2
- MF/3
- G/90

These are all essentially lump sum contracts with some element of contractor's design. MF/1 and G/90 are contracts for supply and erection. MF/2 and MF/3 are supply only contracts. All four contracts are intended principally for new installations rather than for repairs and maintenance.

Publishers of the model forms

MF/1, MF/2 and MF/3 are published by the Institution of Electrical Engineers on behalf of the Joint IMechE/IEE Committee on Model Forms of General Conditions of Contract. The joint committee is widely representative of the industry and it includes suppliers and users of electrical and mechanical plant together with engineers and lawyers.

G/90 is published on behalf of the Water Services Association (WSA) for water industry lump sum plant contracts. The form is registered with the Office of Fair Trading and three trade associations, the Federation of British Electrical and Allied Manufacturers' Associations, the British Pump Manufacturers' Association and the British Effluent and Water Association have confirmed that its conditions are not objectionable.

MF/1

MF/1 was published in 1988 as an entirely new model form suitable for home and export contracts. It was derived from, and replaced, two earlier model forms, MF 'A', which first appeared in 1921, and MF 'B3', which dated from 1954. MF/1 is intended for use in projects covering the supply, delivery and erection on site of electrical and mechanical plant. It contains additional special conditions for contracts involving incidental supply of hardware and software. The contract is administered by an engineer who is required to exercise his discretion fairly between the parties and it contains extensive provisions for on-site testing both before and after taking-over. Additional special conditions for use in contracts where payments are based on measurement of the works were published in 1993 under the title Supplement No. 1 (S1-MF/1). MF/1 is recommended for use by the Association of Consulting Engineers.

MF/2

MF/2 was published in 1991 as a new model form for home or overseas contracts for the supply of electrical and mechanical plant. It was derived from, and replaced, two earlier model forms, MF 'B1', which originated in 1925, and MF 'B2', which dated from 1928. MF/2 is intended for use on projects similar to those on which MF/1 might be used but the essential difference is that erection of the plant on site is performed by persons other than the supplier (who is nevertheless referred to as the contractor). The provisions of MF/2 follow as closely as practicable the provisions of MF/1 and the text is identical in many clauses. The contract is administered by an engineer notwithstanding the omission of on-site obligations of the parties. It is the omission of these obligations, particularly those relating to tests on completion and performance tests, which marks the main difference between MF/2 and MF/1.

To cover the possibility that the supplier of the plant may be required to provide supervisory services during erection, MF/2 contains a supplementary model form entitled 'Form of Supervision Contract'. Its use is, of course, optional. MF/2, like MF/1, is recommended for use by the Association of Consulting Engineers.

MF/3

MF/3 was published in 1993 and it derives from an earlier model form, Model Form 'C', which was available in various editions between 1924 and 1993. MF/3, like MF/2, is a supply only contract but it is a much simpler and shorter model form. It is suited more to the procurement of electrical and mechanical goods than to plant. It is not an engineer administered contract and the parties are known as the vendor and the purchaser.

G/90

G/90, as its title implies, was published in 1990. It is basically a repro-
duction of the 1976 edition of Model Form 'A' with 1978 and 1982
amendments further amended by:

- Form G amendments published in 1978;
- Additional and substitute clauses recommended by WSA in 1989.

G/90 retains the style and much of the text of the old MF 'A' but its
various amendments are intended to give the purchaser more control
over the contractor and greater protection for latent defects.

The contract is intended, as is MF/1, for the supply and erection of
electrical and mechanical plant but it is particularly aimed at, and used
for, plant contracts in the water industry. Like MF/1 it is engineer
administrated and it contains provisions for on-site testing both before
and after taking-over. It is not intended for overseas use.

1.2 MF/1 – Revision 3 (1995)

The commentary in this book is based on the 1988 edition of MF/1 as
reprinted in 1992 with the 1989 and 1992 editorial amendments. The
extracts from MF/1 are taken from the 1992 reprint.

In 1995 the publishers of MF/1 intend to issue a revised version of the
model form containing, I am pleased to say, some amendments found
necessary as a result of my work on this book. This 1995 version of MF/1
will, I am informed, be known as MF/1 Revision 3.

By kind consent of the publishers of MF/1 I am able to include in this
book (section 2.2) a summary of the draft changes intended for Revision 3
and I have been able to indicate in the text of the book the impact of the
changes.

1.3 Other standard forms

Whilst the IMechE/IEE model forms, along with G/90, are undoubtedly
the principal standard forms of contract for electrical and mechanical
plant a brief mention should be made here of certain other standard forms
which I have not attempted to cover in any detail in this book:

- the BEAMA Conditions for the reconstruction, modification or repair of
  plant and equipment in the UK involving work on site. BEAMA is the
  Federation of British Electrotechnical and Allied Manufacturers'
  Associations;

- the FIDIC Conditions of Contract for electrical and mechanical works including erection on site. FIDIC is the Fédération Internationale Des Ingénieurs-Conseils;
- the IChemE Conditions of Contract for process plant. The Red Book is a lump sum contract; the Green Book is a cost-reimbursable contract.

1.4 Benefits of model forms

Some suppliers of plant and some purchasers prefer to use their own in-house forms of contract rather than model forms – presumably on the basis that their interests are best served by doing business on their own terms. Whether in the long term this is a sound proposition is debatable.

There is a theory that every project has an inherent gross cost which is shared between the parties according to the terms of the contract and the allocation of risk. The lowest gross cost is thought to occur when the risks are shared equitably with each party carrying the risks he is best able to control. A contract on these terms it is argued is more likely to be in the best interests of both parties than a biased contract which looks only to the interests of one party.

Assuming this to be true, the primary benefits of using model forms produced on a consensus basis are that such forms are cost effective and they provide the foundation for the contract to be performed in an amicable manner.

Secondary benefits of using model forms are:

- development of expertise by repeat use and experience;
- familiarity with the contents of the model forms;
- consistency of application of contractual provisions;
- the availability of professional training on the model forms;
- regular review and up-dating by a competent drafting committee.

There is then the further point that many businesses are fully occupied in coping with the technical and financial aspects of their operations and it may be neither good use of resources nor good policy to take on the burden of producing *ad hoc* or in-house forms of contract.

1.5 Characteristics of the model forms for plant contracts

Whatever might be said by way of criticism about the drafting of certain clauses of the model forms for plant contracts later in this book one thing which has to be recognized is that all these forms have a remarkably dispute-free history of use.

The number of cases which have reached the courts on the current forms or their predecessors is minimal and arbitrations are believed to run at a level which is a mere fraction of that in the dispute-prone construction industry. Clearly the model forms have some characteristics which might prove a lesson to others. Perhaps the most significant characteristics in this respect are those relating to limitation of liabilities and exclusive remedies.

Limitations of liability

At first sight limitations of liability, which cap the financial liability of the contractor and restrict his liability for defects to 12 months or some other short period, might appear as unfair contract terms. Perhaps in some cases they are. But generally they are necessary to protect the contractor from a scale of risk which, if left open-ended, would destroy the commercial sense of being in the plant supply business.

With regard to defects and latent damage there is the valid point that if moving machinery is going to develop defects, then such defects are far more likely to be revealed during the defects liability period than after the period has ended. Which is, of course, the reverse of the position in the construction industry where latent damage frequently comes to light after the defects liability period has ended.

A detailed examination of the financial and time-related limitations of liability in the model forms is made in Chapter 14.

Exclusive remedies

The provisions in MF/1 and MF/2 that the conditions of contract are exhaustive as to the rights, obligations and liabilities of the parties are much discussed in this book.

If it is the case that the effect of the exclusive remedy provisions is wholly to exclude implied terms and legal actions brought on implied terms then it does mean that when the parties sign the contract they know exactly what they are letting themselves in for. What they see is what they get. This is potentially bad news for litigants who seek to rely on implied terms or at least plead them in the alternative.

But one consequence of exclusive remedy provisions is that the parties to the contract must give more attention to the express terms of the contract than they might otherwise regard as necessary.

1.6 Amendments to the model forms

Some users of the model forms do, as a matter of policy, amend the forms – usually to bring them closer into line with the provisions they are familiar with in other standard forms. As might be expected limitation of liability clauses and exclusive remedy provisions are frequently top of the list for such amendment. Other common amendments relate to clarification of design responsibility and to the obligations of the parties on site-related matters.

The commentary which follows in later chapters on the various clauses of the model forms reveals those clauses most likely to be the subject of amendment.

1.7 Interpretation of the model forms

The point is made above that there have been very few legal decisions on the model forms for plant contracts. There is, therefore, little by way of judicial guidance on their interpretation.

In writing this book I have had to rely on my own opinion and such decisions as I consider relevant from construction cases. At times it may seem that I have raised more questions than I have answered. But if in doing so I have drawn attention to difficult points of interpretation and alerted users of the model forms to the need to take legal advice on matters of importance I will, in part, have succeeded in my task.

And that task was to write a book which would assist users of the model forms to see the forms in the context of the law generally and to offer a view on how they should be applied and interpreted.

Chapter 2

Overview of the model forms

2.1 Introduction

The current range of model forms for plant contracts gives users a set of modern and well-presented documents designed to cover procurement requirements over a wide field of activity.

There should be little difficulty in selecting the appropriate form when the choice is being made between MF/1, MF/2 and MF/3 since each clearly has its own function. There may be a point in some projects when the supply of plant under MF/2 and the supply of goods under MF/3 is not far apart in financial or practical terms and then it is a matter of individual procurement policy whether to trade on the detailed provisions of MF/2 or the simpler provisions of MF/3.

MF/1 or G/90

As to the choice between MF/1 and G/90 the problem is that both are intended to serve broadly the same purpose. There is nothing in MF/1 to preclude its use in the water industry and there is nothing in G/90 to restrict its use to the water industry. In fact the water industries of Scotland and Northern Ireland make far more use of MF/1 than of G/90.

G/90 does place particular emphasis on delayed delivery of plant to site – which is certainly a point to consider when there are associated building or civil engineering contracts which may interrupt the planned progress of the plant contract.

But against that, G/90 suffers the disadvantage that it is a difficult contract to operate due to the absence of a contents list, the absence of section headings and the general lack of orderliness in the arrangement of its provisions. These would present a more serious obstacle to the use of G/90 than they actually do were it not for the fact that G/90 is derived from the similarly drafted and arranged old model form MF 'A' that gained widespread familiarity in use over many years and even now still has its supporters.

However, when it comes to examination of the detailed provisions of MF/1 and G/90 it can be seen that there is more of a difference between the two forms than mere style and presentation. MF/1 is highly protective of the contractor's position by a variety of financial and time-related limitations of liability. And it contains a very powerful exclusive remedy clause whereby the rights, obligations and liabilities of the parties are confined to those expressly stated in the contract.

By contrast, G/90 is far more open on such matters and accordingly some argue that it offers a better deal to purchasers than MF/1.

Clearly in putting the contractor under three year latent defects liability it does this but, of course, this particular matter can easily be balanced out in MF/1, and frequently is, by amendment to the same effect. As for other matters the 'pros and cons' are more debatable. Nevertheless it is understood that G/90 came to be produced as a separate model form because the feeling in the English water industry was that the amendments necessary to MF/1 to make it acceptable were too numerous to make the exercise worthwhile.

This has not been the view of the water industry in Scotland and Northern Ireland and the amendments to MF/1, some of which are listed later in this chapter, are not over-extensive.

But putting aside questions of whether MF/1 or G/90 is more favourable to the interests of one party than the other the great advantage of MF/1 is that not only is it the result of a comprehensive modern review of contractual provisions but it is also subject to regular review by a standing committee.

2.2 Model form – MF/1

The commentary in this book is addressed principally to the provisions of MF/1 and as with any other model form of such length and detail it is always possible to find points of weakness in the drafting. However, that should not detract from the strengths of MF/1 which are considerable.

Practical matters

Perhaps the most impressive aspect of MF/1 is the clear and comprehensive way it deals with the practical side of supply, delivery and erection of plant. And it sets out its provisions in a logical sequence mirroring the contractor's progress. Thus:

- obligations of the parties
- inspection and testing before delivery
- suspension of work or deliveries
- tests on completion

- taking-over
- performance tests.

Consequently users of MF/1 should have little difficulty in being able to rapidly identify and understand the provisions applicable to construction of the works. If there is a point of weakness in this area it is that use of the plant by the purchaser before taking-over is not contemplated.

Financial matters

An obvious omission from MF/1, which will have to be corrected in the 1995 Revision 3, is that it contains no reference to value added tax either in its clauses or in the standard forms of tender or agreement. But otherwise the payment provisions which are only intended as fall-back clauses in the absence of special conditions are more than adequate as are entitlements and procedures on claims for additional payments.

Care does need to be taken, however, with one of the major claims clauses in the contract which can easily be overlooked as to its application and effect. This is clause 25.1 which goes under the heading of suspension. The clause however also covers any delay in delivery or erection of the plant caused by the purchaser or any contractor for which he is responsible. And that delay is not only deemed to be a suspension but the contractor, far from being required to mitigate his loss, is expressly required to maintain his resources on or near the site.

Contractual matters

The subject of responsibility for design is discussed in much detail later in this book. It cannot be said that MF/1 does not address the issue but there are some points of uncertainty on what is intended.

There is something of a gap in MF/1 in relation to sectional completions. Surprisingly the provisions for extension of time and damages for delay apply only to the whole of the works. This will certainly be dealt with in the 1995 Revision 3.

As to limitations of liability and the exclusive remedy provisions in MF/1, these also are considered in much detail later. All that can be said here by way of summary is that limitations of liability are so important to both contractors and purchasers that each and every contract let under MF/1 deserves individual attention.

Whether or not the exclusive remedy provisions favour one party or the other or whether they should be left in the contract at all is a matter which could be argued at length. This is certainly one for the lawyers.

Amendments to MF/1

The following list gives some of the more common amendments made
by purchasers to bring MF/1 into line with their particular requirements:

- Clause 1.1.d – the engineer
  - definition altered so that the engineer is any person appointed from
    time to time
- Clause 2.8 – replacement of the engineer
  - deleted to avoid the contractor's control over replacement of the
    engineer
- Clause 4.1 – precedence of documents
  - words added at the end of the clause of the kind 'except for variations
    and instructions duly notified by the Engineer' to clarify a point on
    precedence
- Clause 5.6 – contractor's responsibility for PC items
  - deleted to avoid the contractor having no responsibility for work
    done under provisions sums and prime cost items
- Clause 6.2 – labour, materials and transport
  - deleted to put the risk of rises and falls in prices on the contractor
- Clause 8.2 – failure to provide bond or guarantee
  - final sentence deleted to remove the limitation of the contractor's
    liability
- Clause 11.2 – wayleaves, consents etc
  - amended so that the purchaser has less responsibility for obtaining
    certain consents, wayleaves and approvals (usually accompanied by
    an additional clause 13.4 stating the contractor's obligations in these
    matters)
- Clause 11.5 – purchaser's lifting equipment
  - amended so that the purchaser allows the contractor to use lifting
    equipment available on the site
- Clause 13.3 – contractor's design
  - amended so that the contractor's entitlement to disclaim responsi-
    bility for design provided by the purchaser or the engineer is
    removed
- Clause 15.7 – purchaser's use of drawings
  - amended to give the purchaser greater freedom in the use of draw-
    ings and information supplied by the contractor
- Clause 16.2 – errors in drawings
  - amended to attempt greater compatibility with the contractor's
    responsibility for design under clause 13.3
- Clause 18.4 – opportunities for other contractors
  - amended so that the contractor is obliged to afford reasonable
    opportunities for others within his price and additional payment

only becomes due for affording opportunities as instructed by the engineer which could not have been foreseen at the time of tender
- Clause 19.2 – work at night or on rest days
  - amended so that if the contractor works at night or on rest days to suit his convenience or meet the time for completion the extra costs incurred by the purchaser are borne by the contractor
- Clauses 21.2, 21.3, 21.4 – special loads and extraordinary traffic
  - deleted so the contractor is fully responsible for the costs in moving plant to the site
- Clause 33.1 – extension of time for completion
  - amended to include for sections
- Clause 33.2 – delays by sub-contractors
  - deleted as being superfluous and potentially misleading
- Clauses 34.1 and 34.2 – delay in completion
  - amended to include for sections
- Clause 36.9 – limitation of liability for defects
  - deleted to remove the limitations of the contractor's liability
- Clause 36.10 – latent defects
  - amended so that the contractor is responsible for making good defects which appear within three years (sometimes six years) of taking-over. The proviso of gross misconduct which qualifies the contractor's liability is frequently deleted
- Clause 44.3 – limitation of the contractor's liability
  - deleted or amended to suit
- Clause 44.4 – exclusive remedies
  - deleted to avoid the confinement of rights, obligations and liabilities to those expressly stated in the contract
- Clause 51.1 – applicable law
  - amended to the Laws of Scotland or the Laws of Northern Ireland
- Additional clauses:
  - VAT, contract price exclusive of
  - gifts, inducements and rewards
  - recovery of sums due from the contractor.

1995 Revision 3 to MF/1

The 1995 Revision 3 to MF/1 is expected to contain in the order of 200 changes to the text of MF/1 as presently found in the 1992 reprint.

The great majority of the changes will be points of detail introduced to achieve consistency of terms, consistent use of capitals or grammatical improvement. There will however be some changes of considerable contractual significance and others which, although perhaps less obviously important, are necessary to rectify errors or omissions in the present text.

The following list gives some of the more important changes which it is understood are likely to feature in the 1995 Revision 3:

- Clause 1.1.c – sub-contractor
  - requirement for consent to be in writing may be deleted
- Clause 5.2 – site data
  - the data on which the tender is deemed to be based may be qualified to that which is made available in writing
- Clause 5.6 – prime cost items
  - clarification likely that the exclusion of the contractor's responsibility applies only to work done under prime cost items
- Clause 11.3 – import permits, licences etc
  - the contractor's entitlement to recover additional costs in the event of the purchaser's default may be expressly stated
- Clause 15.6 – operating and maintenance instructions
  - clarification likely that the times for supply may be as stated in the contract not just as in the programme
- Clause 23.3 – services for tests and inspection
  - provisions for tests and inspections on the premises of sub-contractors may be deleted
- Clause 25.1 – instructions to suspend
  - power to suspend likely to be extended to the whole or any part of the works
- Clause 26.1 – defects before taking-over
  - entitlements may be expressly stated to be without prejudice to the purchaser's rights under clause 23.5
- Clause 27.2 – engineer's power to vary
  - likely to be expressed that the engineer alone shall have power to vary the works until taking-over
- Clause 27.3 – valuation of variations
  - clarification probable that it is the engineer who determines the value of variations
- Clause 28.5 – consequences of failure to pass tests on completion
  - point likely to be made that it is the purchaser who is entitled to terminate for default not the engineer
- Clause 29.3 – effect of taking-over certificate
  - correction probable so that risk passes to the purchaser with effect from the date of taking-over as stated in the taking-over certificate
- Clause 35.4 – adjustments and modifications
  - provision likely to be made for additional cost incurred by the purchaser to be deducted from the contract price
- Clause 35.8 – consequences of failure to pass performance tests
  - respective roles of the engineer and the purchaser likely to be clarified
- Clause 37.1 – ownership of plant
  - contract value likely to be corrected to value

- Clause 40.3 – remedies on failure to certify or make payment
  - contractor's entitlement to terminate will probably be extended by removing the reference in clause 40.3(b) to the engineer's failure to certify
- Clause 41.2 – allowance for profit on claims
  - list likely to be extended to include clause 25.6 (effect of suspension on defects liability)
- Clause 43.4 – injury to persons and property
  - the purchaser's indemnity to the contractor may be reworded by reference to the purchaser's risks
- Clause 47.1 – insurance of works
  - clarification may be added in respect of termination
- Clause 49.2 – valuation at date of termination
  - the valuation to be made by the engineer will probably be expressed as the value of the part of the works executed prior to termination
- Clause 54.1 – applicable law
  - by reference to the special conditions the applicable law of the contract may be stated to be other than English law
- Additional special conditions for sectional completion
  - revised clauses 33.1, 33.2 and 34.1 likely to be added.

2.3 Model form – MF/2

Most of the earlier comments on MF/1 apply to MF/2 since the text of the two contracts is identical so far as it is appropriate for it to be so. Indeed many of the 1995 Revision 3 amendments to MF/1 will simply bring minor points of drafting detail into line with the more recently produced MF/2.

Role of the engineer

One aspect of MF/2 which does attract some attention is whether the role of the engineer should follow that in MF/1 as closely as it does. It can be argued that the engineer's role in a supply only contract should be on a different scale to that in a supply and erect contract.

But if it is accepted that contracts let under MF/2 will usually be for the same type of plant as in MF/1 contracts – but for a variety of reasons it is not intended that the supplier should erect the plant – then the case for a contract administrator/supervisor with a duty to act fairly between the parties is more apparent.

2.4 Model form – MF/3

MF/3 as a simple supply of goods contract avoids the detail and complexity of the other model forms for plant contracts.

It might well be said that the proportion of MF/3 taken up by limitations on the vendor's liability for various defaults is out of scale and that the purchaser is not particularly well served by the form. But that begs the question – on whose terms of business would a contract for the supply of goods normally be made?

The answer, of course, is the supplier, (vendor) rather than the purchaser. And in that context it can be argued that MF/3 does give a fair deal to both the vendor and the purchaser.

2.5 Model form – G/90

G/90, as has been stated earlier, is a difficult model form to operate in that its style and layout lacks order. It is easy to miss key provisions in the form but it is not particularly easy to spot what is missing from its provisions.

Perhaps the most notable omission is the absence of any reference to the design obligations of the contractor. Not far behind is the absence of any reference to which party is responsible for obtaining statutory consents, wayleaves and the like. Also conspicuously absent is the mention of any express role for the engineer in the process of valuation and the lack of provisions for termination in the event of prolonged suspension.

Like MF/1, the most important claims provisions of G/90 are tucked away obscurely into the suspension of works clause. Unlike MF/1, and perhaps uniquely amongst modern forms of contract, most of the express claims of G/90 are to be valued in like manner to variations rather than on a cost incurred basis.

Chapter 3

Documents, definitions and notices

3.1 Introduction

This chapter examines the various documents associated with the model forms for plant contracts; the definitions within the model forms; and provisions on the giving of notices.

Contract documents

Not all contracts identify with clarity the documents which are to be taken as binding on the parties and which can rightly be termed contract documents. G/90 for example simply refers to all documents to which reference may properly be made – and this clearly leaves scope for argument on whether particular documents come into this category.

In contrast the IChemE model forms list all the categories of documents and schedules which constitute the contract and then state that no other documents shall be of any contractual effect.

MF/1 and MF/2 both give a definition of the contract which aims for precision in listing:

- the tender
- the letter of acceptance
- the conditions of contract
- the specification
- the drawings as annexed
- the schedules as referred to.

This does not appear as firm as the IChemE model forms which in relation to pre-contractual correspondence and negotiations positively require all documents which are to have contractual effect to be identified and listed in one of the standard categories. But good practice requires that this be done in any event.

Precedence of documents

There is a general rule of legal construction that the particular prevails over the general but beyond this there is no general rule giving precedence to one class of documents over another class. If such precedence is to exist it has to be stated in the contract itself.

Some contracts such as the ICE Conditions leave the matter of precedence completely open but other contracts, including MF/1 and MF/2 do address the issue.

In MF/1 and MF/2 the order of precedence is expressly stated to be:

- the conditions as amended by the letter of acceptance – with the special conditions prevailing over the general conditions;
- the specification;
- other contract documents.

The IChemE model forms have similar rules but go further in stating that the description of works in schedule 1 prevails over the specification.

G/90 does not set any order of precedence.

Associated documents

Within or accompanying most model forms of contract are documents such as specimen forms of bonds, guarantees and the like; specimen forms of tender and agreement; model forms of sub-contract; and specimen certificates.

These will normally only be contract documents where they formalize obligations and responsibilities of the parties as, for example, in the forms of tender and agreement. But if their use is made mandatory in the contract, for example, when a particular bond is to be entered into on a standard form of bond, then failure to use the relevant document could be a breach of contract.

A schedule of the various documents included within the model forms for plant contracts is given in the next section of this chapter.

Definitions

Most model forms of contract commence with a schedule of definitions of key terms to assist in interpretation of the contract. Usually the defined terms are identified in the text of the contract by capital letters or bold print to alert users of the contract to their existence as defined terms. Plant contracts tend to use capitals; process contracts tend to use bold print.

There are some obvious areas of similarity between contracts on which terms are defined and how they are defined but regrettably there is no reliable common approach. It is therefore unwise to assume that the defined meaning of a term in one contract applies in another contract.

As a broad rule, definitions cover such matters as:

- the parties and the engineer
- the contract documents
- the site and the works
- time-related terms
- financial terms.

Interpretation

The majority of model forms including MF/1, MF/2 and G/90 contain definitions to the effect:

- writing means any handwritten, typed or printed statement;
- words importing persons include firms and organizations with legal capacity;
- words importing the singular include the plural and vice versa.

A further common provision is that headings and marginal notes are not to be taken into consideration in construction of the contract. Note that in G/90 this appears not with the definitions but in clause 38.

Applicable law

Not all model forms of contract make a statement as to the applicable law. MF/1, MF/2 and G/90 however all contain provisions that the contract should be governed and interpreted in accordance with English law. MF/3 provides the option of Scottish law or Northern Ireland law.

It is understood that the 1995 Revision 3 to MF/1 will indicate that where the applicable law is to be other than English law that should be stated in the special conditions.

Notices

Provisions on notices vary from simple statements that notices shall be in writing to comprehensive requirements detailing:

- the form of notices
- the method of delivery

- the addresses for delivery
- the time of service.

MF/1 and MF/2 both follow the comprehensive approach; MF/3 and G/90 are less precise.

3.2 Contents of the model forms (see Table 3.1)

Table 3.1 Contents of the model forms

	MF/1	MF/2	MF/3	G/90
Introductory notes	●	●	●	●
Table of contents	●	●	●	
List of definitions	●	●		●
General conditions	●	●	●	●
Appendix to the conditions	●	●	●	●
Notes on special conditions	●	●		
Special conditions for hardware and software	●			
Form of sub-contract	●	●		
Form of supervision contract		●		
Explanatory notes	*	*		●
Form of tender	●	●	●	
Form of agreement	●	●	●	
Form of performance bond	●	●	●	●
Retention repayment guarantee				●
Bank repayment bond				●
Insurance company repayment bond				●
Parent company guarantee				●
Variation of price formula				●
Variation order form	●	●		
Taking-over certificate	●			
Index	●	●	●	●
Supplement for remeasurement	**			

*   published in separate documents entitled *Commentary on MF/1 (MF/2) – a practical guide*. The MF/2 Commentary includes a specimen defects liability bond.
**   available as a separate publication.

3.3 Documents, definitions and notices in MF/1 and MF/2

In this section the following clauses of MF/1 and MF/2 are considered:

- clauses 1.1 to 1.5   – definitions and interpretations
- clause 4.1   – precedence of documents

- clause 7.1 – agreement
- clause 8.1 – provision of performance bond or guarantee
- clause 8.2 – failure to effect bond or guarantee
- clause 9.1 – details confidential
- clauses 10.1 to 10.3 – notices.

With the exception that not all the definitions in MF/1 are included in MF/2 (e.g. tests on completion and performance tests are omitted) the above clauses are the same in numbering and wording in MF/1 and MF/2). The following clauses are all taken from MF/1.

Clause 1.1.a – 'Purchaser'

'Purchaser' means the person named as such in the Special Conditions and the legal successors in title to the Purchaser but not (except with the consent of the Contractor) any assignee of the Purchaser.

It may seem unnecessary for the purchaser to be formally named in the special conditions but this is a worthwhile precaution against uncertainty as to the exact legal identity of the purchaser in the event of dispute.

For detailed comments on assignment see Chapter 7. It is, however, difficult to envisage circumstances where the contractor would consent to assignment of the purchaser's obligation to pay the contract price without a formal novation of the contract.

Clause 1.1.b – 'Contractor'

'Contractor' means the tenderer whose Tender has been accepted by the Purchaser and the legal successors in title to the Contractor but not any assignee of the Contractor.

There is no specified requirement that the contractor should be named in the special conditions although the *aide mémoire* to the special conditions provides a space under clause 10.2 for the contractor's address.

Clause 1.1.c – 'Sub-Contractor'

'Sub-Contractor' means any person (other than the Contractor) named in the Contract for any part of the Works or any person to whom any part of the Contract has been sub-let with the consent in writing of the Engineer, and the Sub-Contractor's legal successors in title, but not any assignee of the Sub-Contractor.

It is questionable whether sub-contractor should be a defined term. There is a danger that it creates a contractual divide between the contractor and the sub-contractor which is not intended. For further comment see Chapter 7.

Clause 3.2 (sub-contracting) does not expressly require the consent of the engineer to sub-contracting to be given in writing and it is understood that the 1995 Revision 3 to MF/1 will delete the words 'in writing' from clause 1.1.c.

Clause 1.1.d – 'Engineer'

'Engineer' means the person appointed by the Purchaser to act as Engineer for the purposes of the Contract and designated as such in the Special Conditions or, in default of any appointment, the Purchaser.

The contractor is entitled to know the identity of the engineer when he tenders and there is no provision for the engineer to be changed without the consent of the contractor. See the comment on clause 2.8 in Chapter 8.

Clause 1.1.d – 'Engineer's Representative'

'Engineer's Representative' means any assistant of the Engineer appointed from time to time to perform the duties delegated to him under Clause 2 (Engineer and Engineer's Representative) hereof.

It is questionable whether there can be more than one engineer's representative. What is not in doubt is that the engineer's representative can be changed without reference to the contractor notwithstanding the fact that he may have exercise of the full range of delegated duties.

Clause 1.1.f – 'The Conditions'

'The Conditions' means these General Conditions and the Special Conditions.

This is a useful definition in clarifying a point which might otherwise be contentious.

Clause 1.1.g – 'Contract'

'Contract' means the agreement between the Purchaser and the Contractor (howsoever made) for the execution of the Works including the Letter of Acceptance, the Conditions, Specification and the drawings (if any) annexed thereto and such schedules as are referred to therein and the Tender.

The words 'howsoever made' indicate that execution of a formal agreement is not essential.

It is to some extent a matter of preference into which of the categories of documents any particulars negotiated or agreed between the parties beyond those in the tender documents are incorporated. But what is essential is that they should be formally incorporated somewhere in the documents.

Clause 1.1.h – 'Contract Price'

'Contract Price' means the sum stated in the Contract as the price payable to the Contractor for the execution of the Works.

In MF/2 this definition continues 'adjusted to give effect to such additions and deductions as may be made in accordance with Clause 21 (variations)'.

The difficulty with the definition as it stands in MF/1 is that the sum stated in the contract is not necessarily the price payable to the contractor. For further comment see Chapter 18.

Clause 1.1.i – 'Contract Value'

'Contract Value' means such part of the Contract Price, adjusted to give effect to such additions or deductions as are provided for in the Contract, other than under Sub-Clause 6.2 (Labour, Materials and Transport), as is properly apportionable to the Plant or work in question. In determining Contract Value the state, condition and topographical location of the Plant, the amount of work done and all other relevant circumstances shall be taken into account.

The contract value is something which comes into effect only when there is suspension, termination, or payment of damages for late completion.

Clause 1.1.j – 'Cost'

'Cost' means all expenses and costs incurred including overhead and financing charges properly allocable thereto with no allowance for profit.

This definition is discussed in detail in Chapter 17. The point to make here is that the contractor's entitlement to additional payment is expressed in terms of 'cost' and not 'loss and expense' which is a potentially wider concept.

Clause 1.1.k – 'Tender'

'Tender' means the Contractor's priced offer to the Purchaser for the execution of the Works.

The price submitted in the form of tender may, of course, be adjusted before acceptance. This definition it is suggested should apply to the final offer price but see the definition of 'letter of acceptance' below.

Clause 1.1.l – 'Letter of Acceptance'

'Letter of Acceptance' means the formal acceptance by the Purchaser of the Tender incorporating any amendments or variations to the Tender agreed by the Purchaser and Contractor.

There is a danger in this arrangement of the letter of acceptance becoming a counter offer. It is far better to have a clean acceptance of a written offer (revised if appropriate) from the contractor than to attempt to incorporate a revised offer in the letter of acceptance.

Clause 1.1.m – 'Time for Completion'

'Time for Completion' means the period of time for completion of the Works or any Section thereof as stated in the Contract or as extended under Sub-Clause 33.1 (Extension of Time for Completion) calculated from whichever is the later of:

(a) the date specified in the Contract as the date for commencement of the Works;
(b) the date of receipt of such payment in advance of the commencement of the Works as may be specified in the Contract;
(c) the date any necessary legal, financial or administrative requirements specified in the Contract as conditions precedent to commencement have been fulfilled.

The purpose of this definition is to establish the date from which the time for completion starts to run. The stipulated time for completion needs to be entered in the special conditions or a contract schedule of timing requirements.

Clause 1.1.n – 'Contractor's Equipment'

'Contractor's Equipment' means all appliances or things of whatsoever nature required for the purposes of the Works but does not include Plant, materials or other things intended to form or forming part of the Works.

Clause 1.1.o – 'Plant'

'Plant' means machinery, computer hardware and software, apparatus, materials, articles and things of all kinds to be provided under the Contract other than Contractor's Equipment.

Taken together these definitions of equipment and plant provide a clear distinction between things used to construct the works and things incorporated into the works.

Clause 1.1.p – 'Works'

'Works' means all Plant to be provided and work to be done by the Contractor under the Contract.

No distinction is made here between permanent and temporary works and the defined term probably covers both.

Clause 1.1.q – 'Section of the Works'

'Section of the Works' or 'Section' means the parts into which the Works are divided by the Specification.

A 'section' is normally distinguished from a 'part' by having its own specified time for completion.

Clause 1.1.r – 'Programme'

'Programme' means the programme referred to in Clause 14 (Programme).

This programme which is submitted after acceptance of the tender will rarely be a contract document. But see Chapter 11 for further comment.

Clause 1.1.s – 'Specification'

'Specification' means the specification of the Works annexed to or included in the Contract including any modifications thereof made under Clause 27 (Variations).

The purpose of this definition is to emphasize that the specification may be varied and that the revised specification then becomes the contract specification.

Clause 1.1.t – 'Special Conditions'

The 'Special Conditions' means the alterations to these General Conditions specified and identified as the Special Conditions in the Contract.

This is slightly misleading because special conditions frequently contain additions to rather than alterations to the general conditions.

Clause 1.1.u – 'Site'

'Site' means the actual place or places, provided or made available by the Purchaser, to which Plant is to be delivered or where work is to be done by the Contractor, together with so much of the area surrounding the same as the Contractor shall with the consent of the Purchaser actually use in connection with the Works otherwise than merely for the purposes of access.

It is suggested that from this definition the site is only the place provided by the purchaser and if the contractor arranges to occupy adjacent places with the consent of others that is not part of the site.

Clause 1.1.v – 'Tests on Completion'

'Tests on Completion' means the tests specified in the Contract (or otherwise agreed by the Purchaser and the Contractor) which are to be made by the Contractor upon completion of erection and/or installation before the Works are taken over by the Purchaser.

This definition makes the important point that tests on completion precede taking-over.

Clause 1.1.w – 'Performance Tests'

'Performance Tests' means the tests (if any) detailed in the Specification or in a performance test schedule otherwise agreed between the Purchaser and the contractor, to be made after the Works have been taken over to demonstrate the performance of the Works.

The words '(if any)' establish that performance tests apply only where they are particularly required. Performance tests follow taking-over.

Clause 1.1.x – 'Defects Liability Period'

'Defects Liability Period' has the meaning assigned by Sub-Clause 36.1 (Defects Liability).

This is what is known as a circular definition. It simply refers to a definition given elsewhere. See Chapter 14 for further comment.

Clause 1.1.y – 'Purchaser's Risks'

'Purchaser's Risks' has the meaning assigned by Sub-Clause 45.1 (Purchaser's Risks).

These are discussed in Chapter 15. They are matters of great importance in connection with damage, injuries and insurances.

Clause 1.1.z – *'Force Majeure'*

'Force Majeure' has the meaning assigned by Clause 46 (*Force Majeure*).

Force majeure is not a term of English law and consequently it needs to be defined to give it meaning. See Chapter 19 for further comment.

Clause 1.1 – Miscellaneous

1.1.aa	'Appendix' means the Appendix to these General Conditions.
1.1.bb	'writing' means any hand-written, type-written or printed statement.
1.1.bb	'day' means calendar day.
1.1.dd	'week' means any period of 7 days.
1.1.ee	'month' means calendar month.

Note that a 'day' is a calendar day and not a working day.

Clause 1.2 – Interpretation

Words importing persons or parties shall include firms, corporations and any organisation having legal capacity.

This is a necessary definition since both the purchaser and the engineer are defined as persons.

Clause 1.3 – Singular and plural

Words importing the singular only also include the plural and vice versa where the context requires.

The key words here are 'where the context requires'. That is to say it is not always the case that the singular includes the plural.

Clause 1.4 – notices and consents

> Wherever in the Conditions provision is made for the giving of notice or consent by any person, unless otherwise specified such notice or consent shall be in writing and the work 'notify' shall be construed accordingly. Any consent required of a party or the Engineer shall not be unreasonably withheld.

This provision establishes an important contractual point – that 'notify' means giving notice in writing.

The provision that consent shall not be unreasonably withheld raises the difficult question – is it a breach of contract to do so and if it is what remedies, if any, have the parties.

Clause 1.5 – Headings and marginal notes

> The headings or marginal notes in the Conditions shall not be deemed part thereof or be taken into consideration in the interpretation or construction thereof or of the Contract.

Most standard contracts have similar clauses to this probably to avoid the brevity of the headings and marginal notes distorting the intention of the text of the contract. Nevertheless the construction of contracts without reference to headings and marginal notes can sometimes lead to unexpected interpretation of contractual provisions.

Clause 4.1 – precedence of documents

> Unless otherwise provided in the Contract the Conditions as amended by the Letter of Acceptance shall prevail over any other document forming part of the Contract and in the case of conflict between the General Conditions and the Special Conditions the Special Conditions shall prevail. Subject thereto the Specification shall prevail over any other document forming part of the Contract.

The opening words of this clause indicate that the contract may provide its own particular order of precedence.

There is no mention in the clause of schedules and it can be a matter of some importance in relation to precedence as to whether they are made schedules to the conditions or schedules to the specification. If the schedules are not expressly linked to either then they rank below both.

As worded the clause might be said to cast some doubt on the precedence of engineer's instructions, particularly with regard to changes from the original specification but other clauses in the contract make it clear that the contractor must comply with instructions and variations.

Clause 7.1 – agreement

Either party shall be entitled to require the other to enter into an agreement in the form annexed with such modifications as may be necessary within 45 days after the Letter of Acceptance. The expenses of preparing, completing and stamping the agreement shall be borne by the party making such request and he shall provide the other party free of charge with a copy of the agreement.

The usual reason for entering into a formal contract agreement is to execute the contract as a deed thereby extending the period of limitation for bringing an action for breach of contract from six years to twelve years. But see the comments on time-related limitations in Chapter 14 which indicate that an extended limitation period of twelve years serves little purpose in the model forms for plant contracts.

A secondary benefit is that the formal agreement does bring together on one document both offer and acceptance and the agreement can therefore be seen as constituting in a literal sense 'the contract'.

Clause 8.1 – provision of bond or guarantee

If required by the Purchaser the Contractor shall provide the bond or guarantee of an insurance company, a bank or other surety for the due performance of the Contract. Unless otherwise specified in the Special Conditions, the terms of the bond or guarantee shall be in the form annexed to these General Conditions.

Unless otherwise specified in the Contract the Contractor shall provide the bond or guarantee at his own Cost.

The amount of the bond, the period of its validity, the procedure to be followed for its forfeiture, the arrangements for its release and the currency of any monetary transactions involved shall be stated in the Special Conditions.

Bonds differ considerably in their drafting and the conditions under which they can be called in for payment. At one end of the scale there are 'demand' bonds which can be called in without proof of default or proof of loss; at the other end there are performance or 'conditional' bonds which can only be called upon with certification of default and proof of loss. The model form of performance bond in MF/1 is of the latter type.

However, it should be added that the law is currently in a state of uncertainty on bonds in other contracts which were previously thought to be well understood as a result of two cases in 1994:

- *Perar bv* v. *General Surety*, and
- *Trafalgar House* v. *General Surety*.

Bonds therefore are best left to lawyers.

Clause 8.2 – failure to provide bond or guarantee

If the Contractor shall have failed to provide the bond or guarantee within 30 days after the date of the Letter of Acceptance or within such further period as may be advised by the Purchaser, the Purchaser shall be entitled to terminate the Contract by seven days notice to the Contractor. In the event of termination under this Clause the Contractor shall have no liability to the Purchaser other than to repay to the Purchaser all Costs properly incurred by the Purchaser incidental to the obtaining of new tenders.

For comment on this clause see Chapter 19.

Clause 9.1 – details confidential

The Purchaser and the Contractor shall treat the details of the Contract and any information made available in relation thereto as private and confidential and neither of them shall publish or disclose the same or any particulars thereof (save insofar as may be necessary for the purposes of the Contract), without the previous consent of the other provided that nothing in this Clause shall prevent the publication or disclosure of any such information that has come within the public domain otherwise than by breach of this Clause.

As with many other clauses in MF/1 (and MF/2) the question here is does either party have a remedy for breach by the other.

In normal circumstances an action for damages for breach of contract would apply but the exclusive remedies provisions in Clause 44(4) may prevent such an action.

Clause 10.1 – notices to purchaser and engineer

Any notice to be given to the Purchaser or to the Engineer under the Contract shall be served by sending the same by post, telex, cable or facsimile transmission to, or by leaving the same at, the respective addresses nominated for that purpose in the Special Conditions.

Clause 10.2 – notices to contractor

All certificates, notices or decisions, instructions and orders to be given by the Engineer or the Purchaser under the Contract shall be served by sending the same by post, cable, telex or facsimile transmission to, or by leaving the same at, the Contractor's principal place of business or such other address as the Contractor shall nominate for that purpose.

Clause 10.3 – service of notices

Any notice sent by telex, cable or facsimile transmission shall be deemed to have been served at the time of transmission. A notice sent by post shall be deemed to have been served four days after posting.

For important notices electronic transmission is more effective than post in that service is not only more immediate but easier to establish as having occurred at a particular point in time.

3.4 Documents, definitions and notices in MF/3

MF/3 has only brief provisions relating to its documents. It has no list of definitions and no clauses relating to notices.

Clause 1.1 – precedence of documents

These General Conditions shall have effect subject to any express stipulation or condition at variance with these Conditions that may be contained in the Specification or may otherwise be incorporated in the Contract.

This clause means no more than that the general conditions of contract apply unless there are particular conditions to the contrary elsewhere in the contract.

Clause 1.2 – headings and marginal notes

The headings or the marginal notes hereto shall not be deemed part of these General Conditions or be taken into consideration in the interpretation or construction thereof or of the Contract.

3.5 Documents and definitions in G/90

The relevant clauses of G/90 are:

- clause 1 – definitions
- clauses 3(i) and (ii) – security for due performance

- clause 3(iii)   – expenses of agreement
- clause 38   – construction of contract.

There are no provisions on the giving of notices.

Definitions

G/90 contains fewer defined terms than MF/1 but of those listed only the definition of contract is significantly different.

> The 'Contract' shall mean the agreement between the Purchaser and the Contractor for the execution of the Works howsoever made, including therein all documents to which reference may properly be made in order to ascertain the rights and obligations of the parties under the said agreement.

This is a more flexible definition than in MF/1 but, of course, it has the potential of uncertainty as to which documents do actually form the contract.

Security for due performance

Clause 3(i) of G/90 obliges the contractor to provide security for due performance if so required by the purchaser as does clause 8.1 of MF/1. A major difference, however, is that in G/90 the purchaser pays the costs whereas in MF/1 the contractor pays.

Clause 3(ii) of G/90 deals with failure to provide the required bond or guarantee in much the same way as clause 8.2 of MF/1.

Expenses of agreement

> The expenses of preparing, completing and stamping the agreement, if any, shall be paid by the Purchaser, and an executed counterpart thereof properly stamped together with copies of all other documents comprising the Contract shall be furnished to the Contractor free of charge.

Clause 3(iii)

There is no provision here for either party to require the other to enter into a formal agreement. Nor is there any specimen form of agreement in G/90.

Note that the costs of preparing the agreement are borne by the purchaser whereas in MF/1 the costs fall on the party requesting the agreement.

Construction of the contract

Clause 38

The Contract shall in all respects be construed and operate as an English Contract and in conformity with English law, and all payments thereunder shall be made in sterling money. The marginal notes hereto shall not affect the construction hereof.

The provisions here correspond with those in clauses 1.5 and 54.1 of MF/1.

Chapter 4

Design, drawings and patents

4.1 Introduction

This chapter deals principally with design obligations and responsibilities but includes as related topics the submission and approval of drawings and indemnities on patents.

Design

Design can be defined as the accumulation of ideas and details which go into the production of an artefact or the construction of a project.

Three key questions apply to all contracts in which there is some element of design:

- how is the obligation to undertake the design allocated between the parties?
- how is the responsibility for the effectiveness of the design to be allocated between the parties?
- what standard of liability attaches to the party responsible for design? Is it to be skill and care or fitness for purpose?

Design in plant contracts

In plant contracts the answers to these questions would seem to be obvious. On the basis that the contractor is a specialist providing a product for a particular purpose who would doubt that the contractor carries the obligation to design; is responsible for the design; and that the design should be fit for its purpose.

But that as will be seen later in this chapter in some detail is a gross simplification of the position. The contractor may not be obliged to undertake all of the design – part may be undertaken by the engineer. The contractor may not be required to accept responsibility for all of the design – part may remain the responsibility of the purchaser. And the standard of design liability will not necessarily be fitness for purpose.

Contractual provisions on design

It is an odd feature of many standard forms of contract in the plant, process and construction fields that they either fail completely to address the important issues of design obligations, responsibilities and liabilities or they do so in less than certain fashion. Not surprisingly disputes on design are commonplace.

In the recent case of *Shanks and McEwan (Contractors) Ltd* v. *Strathclyde Regional Council* (1994) a contract under the ICE Conditions of Contract Fifth edition included a provision that 'All tunnel and shaft segments ... shall be of approved design and shall be supplied by an approved manufacturer'. After the segments were installed they cracked and remedial works were ordered. The responsibility for the costs of these works fell on the employer because, contrary no doubt to the intentions of the engineer and the employer, the responsibility for design had not passed to the contractor.

That case highlights the point that in some standard forms, including the ICE Fifth edition, responsibility for design rests with the employer/ purchaser unless it is expressly passed to the contractor.

Amongst the plant contracts MF/1 and MF/2 have gone furthest in dealing with design obligations and responsibilities. MF/3 as simply a supply of goods contract needs to say little. G/90 also says little and this is perhaps one of the most serious deficiencies of G/90.

Skill and care/fitness for purpose

The question of whether the standard of liability for design is on a skill and care basis or a fitness for purpose basis goes well beyond the limits of plant and construction contracts. As will be seen later in this chapter under *IBA* v. *EMI* (1980) – a leading case on design liability (and one of the few cases to reach the courts on MF 'A') – much of the judgment is on medical analogy! Consequently it is a matter of regular debate whenever the contractor takes responsibility for design as to the standard of his liability.

The point of this debate is that if the contractor's responsibility for design is limited to skill and care corresponding to that of a professional designer, then negligence must be proved to establish breach of duty – whether in contract or in tort. That is design failure may occur but the contractor will not be liable provided he has used proper skill and care. However, if the contractor's responsibility is fitness for purpose that is a strict responsibility. The contractor may have used all proper skill and care but if the specified contractual objective is not achieved the contractor will be liable to the purchaser for damages.

Generally, therefore, skill and care is a lower standard of responsibility than fitness for purpose. Consequently contracts requiring fitness for purpose are thought to be more onerous on the contractor than skill and care contracts. Additionally they create a problem for the contractor in that it is difficult for him to pass on responsibilities of fitness for purpose to any professional designers which he may employ.

As a general rule contracts for the supply of services are on a skill and care basis and contracts for the supply of goods are on a fitness for purpose basis. But more on this follows.

What can be said here is that plant contracts, by their supply of goods nature, and by the testing regimes which govern the fulfilment of the contractor's obligations, appear to fall notionally into the fitness for purpose category. But when the provisions of the model forms are examined in detail it can be seen that fitness for purpose may be of very limited application. Frequently because the contractor will end up with responsibility for the design of only part of the plant. And because fitness for purpose of a part says nothing about fitness for purpose of the plant.

4.2 *General principles of design liability*

Any designer, whether he be a professional designer or a contractor designer, can have liabilities in contract or in the tort of negligence. The doctrine of privity of contract confines liabilities in contract to the parties to the contract; but in tort liabilities can extend to third parties.

Liability in contract

Contractual liability for design may arise from:

● the express terms of the contract
● terms implied by common law
● terms implied by statute.

This applies to both design and construct contracts and the contracts of engagement of professional designers.

Express terms

Some contracts state not only the contractor's obligations and responsibilities for design but also the standard of design liability.

Thus in the main form of building contract for design and build, JCT 81, the contractor's liability is limited to like liability with a professional designer. A similar level of liability is intended in the ICE Design and

Construct Conditions of Contract which refer to 'reasonable skill, care and diligence'.

In contrast the IChemE Red Book expressly requires the plant to be suitable for the purposes for which it is intended and the ACA/BPF building form is even more definitive in requiring that 'those parts of the Works to be designed by the Contractor will be fit for the purposes for which they are required'.

The model forms for plant contracts do not directly address the contractor's design liability in this way. But their intention it is suggested is fairly clear – that the contractor's liability is on a fitness for purpose basis. And even if it is difficult to identify a single clear provision to this effect there is a general rule of construction of contracts that the intention is to be collected from the whole of the agreement.

Terms implied by common law

The conditions which must be satisfied for a term to be implied into a contract on common law principles were set out by Lord Simon in the case of *BP Refinery Ltd* v. *Shire of Hastings* (1977). They are:

- it must be reasonable and equitable;
- it must be necessary to give business efficacy to the contract, so no terms will be implied if the contract is effective without it;
- it must be so obvious that 'it goes without saying';
- it must be capable of clear expression;
- it must not contradict any express term of the contract.

Important implied terms relating to plant, process and construction contracts are:

- In a contract for labour –
  to do work in a good and workmanlike manner. *Duncan* v. *Blundell* (1820) where it was said:

  'Where a person is employed in a work of skill, the employer buys both his labour and his judgment; he ought not to undertake the work if it cannot succeed, and he should know whether it will or not. Of course it is otherwise if the party employing him choose to supersede the workman's judgment by using his own.'

  Approved in *Young & Marten* v. *McManus Childs* (1969).
- In a contract for labour and materials –
  to supply materials of good quality and reasonably fit for their purpose. *Myers* v. *Brent Cross Service Co* (1934) where it was said:

'A person contracting to do work and supply materials warrants that the materials which he uses will be of good quality and reasonably fit for the purpose for which he is using them, unless the circumstances of the contract are such as to exclude any such warranty.'

Approved in *Young & Marten* v. *McManus Childs* (1969).
- In a contract for a house –
  that it should be reasonably fit for human habitation. *Hancock* v. *Brazier* (1966)
- In a contract for design and construction of a transmission mast –
  that the finished works will be reasonably fit for their intended purpose. *IBA* v. *EMI* (1980).

Terms implied by statute

The statutes with particular relevance to plant, process and construction contracts and the subject of design liability are:

- The Sale of Goods Act 1979
- The Supply of Goods and Services Act 1982.

Also of some relevance is the Unfair Contract Terms Act 1977 which limits the freedom of suppliers to contract out of statutory obligations.

The Sale of Goods Act 1979 imposes into contracts of sale various warranties, including:

- goods supplied shall correspond with their description;
- goods supplied shall be of merchantable quality;
- goods supplied shall be reasonably fit for any purpose made known to the supplier where the purchaser relies on the skill and care of the supplier.

The Supply of Goods and Services Act 1982 applies to contracts which are not solely for the supply of goods but which include services (such as design). The Act provides amongst other things, for the following terms to be implied into contracts:

- that goods will correspond with description;
- that goods will be of merchantable quality;
- that a supplier acting in the course of business will carry out the service with reasonable skill and care;
- that nothing in the Act prejudices any rule of law which imposes a stricter duty than implied by the Act.

4.3 Fitness for purpose

The benefit to the purchaser when the contractor is under fitness for purpose liability can be seen from these words of Judge John Davies in *Viking Grain Storage* v. *T.H. White Installations Ltd* (1985):

> 'The virtue of an implied term of fitness for purpose is that it prescribes a relatively simple and certain standard of liability based on the 'reasonable' fitness of the finished product, irrespective of considerations of fault and of whether its unfitness derived from the quality of work or materials or design.'

Implied terms

The law, as shown earlier, does not normally imply terms of fitness for purpose into contracts for the employment of professional men. This is how Lord Denning explained the position in *Greaves Contractors Ltd* v. *Baynham Meikle & Partners* (1975):

> 'Apply this to the employment of a professional man. The law does not usually imply a warranty that he will achieve the desired result, but only a term that he will use reasonable care and skill. The surgeon does not warrant that he will cure the patient. Nor does the solicitor warrant that he will win the case. But, when a dentist agrees to make a set of false teeth for a patient, there is an implied warranty that they will fit his gums.'

The position of a contractor, however, is significantly different. The law does imply terms of fitness for purpose into contracts where the contractor supplies materials and the purchaser/employer relies on the contractor's judgment in selection – *Myers* v. *Brent Cross Service Company* (1934).

Fitness for purpose in design and construct contracts

From the starting position above the question is – does the law imply a term of fitness for purpose into every design and construct contract? Clearly it will not do so if there is an express term opposing such an implication but otherwise there is strong presumption in favour of the implied term. The following extracts from the judgment of Lord Scarman in *IBA* v. *EMI* (1980) illustrate this:

> 'In the absence of a clear, contractual indication to the contrary, I see no reason why one who in the course of his business contracts to design,

supply and erect a television aerial mast is not under an obligation to
ensure that it is reasonably fit for the purpose for which he knows it is
intended to be used. The Court of Appeal held that this was the contractual obligation in this case, and I agree with them. The critical
question of fact is whether he for whom the mast was designed relied
upon the skill of his supplier (i.e. his or his subcontractor's skill) to
design and supply a mast fit for the known purpose for which it was
required.

Counsel for the appellants, however, submitted that, where a design,
as in this case, requires the exercise of professional skill, the obligation
is no more than to exercise the care and skill of the ordinarily competent
member of the profession. Although it might be negligence today for a
constructional engineer not to realise the danger to a cylindrical mast of
the combined forces of vortex shedding and asymmetric ice loading of
the stays, he submitted that it could not have been negligence before the
collapse of this mast: for the danger was not then appreciated by the
profession. For the purpose of the argument, I will assume (contrary to
my view) that there was no negligence in the design of the mast, in that
the profession was at that time unaware of the danger. However, I do
not accept that the design obligation of the supplier of an article is to be
equated with the obligation of a professional man in the practice of his
profession. In *Samuels* v. *Davis* [1943] KB 526, the Court of Appeal held
that, where a dentist undertakes for reward to make a denture for a
patient, it is an implied term of the contract that the denture will be
reasonably fit for its intended purpose. I would quote two passages
from the judgment of du Parcq L.J. At p. 529 he said (omitting immaterial words):

> "... if someone goes to a professional man ... and says: 'Will you
> make me something which will fit a particular part of my body?' ...
> and the professional gentleman says: 'Yes', without qualification, he
> is then warranting that when he has made the article it will fit the part
> of the body in question."

And at p. 530 he added:

> "If a dentist takes out a tooth or a surgeon removes an appendix, he is
> bound to take reasonable care and to show such skill as may be
> expected from a qualified practitioner. The case is entirely different
> where a chattel is ultimately to be delivered."

I believe the distinction drawn by du Parcq L.J. to be a sound one. In the
absence of any terms (express or to be implied) negativing the obligation, one who contracts to design an article for a purpose made known
to him undertakes that the design is reasonably fit for the purpose.'

Defence to fitness for purpose

Although the use of reasonable skill and care will not serve as a defence against failure to achieve fitness for purpose, there are other possible defences.

First, amongst these is the defence that the purpose has not been clearly identified in the contract. Second, is the defence, that the works have been used for something other than the intended purpose as stated in the contract.

Plant contracts anticipate this to some extent. In MF/1 the contractor does not warrant that the works when incorporated into some larger project will satisfy the purchaser's requirements. And in G/90 compliance with the specification is sufficient to cause any part of the works to be considered fit for the purpose for which it was intended.

This does raise the prospect that in some circumstances the designer may find the liability of fitness for purpose, even where it expressly applies, easier to dispose of than the liability of using reasonable skill and care. This is because the skill and care liability applies to both the information made available in the tender and other non-contractual information the designer should reasonably have known about. And the designer is not excused the obligation to use skill and care because the purpose of the works is uncertain or because there has been misuse.

4.4 Skill and care

The duty of a professional man to exercise skill and care has long been a legal obligation. In *Lanphier* v. *Phipos* (1838) it was said: 'Every person who enters in to a learned profession undertakes to bring to the exercise of it a reasonable degree of care and skill'.

No better recent exposition will be found than that of Lord Justice Bingham in the case of *Eckersley* v. *Binnie & Partners* (1988). He said:

'The law requires of a professional man that he live up in practice to the standard of the ordinary skilled man exercising and professing to have the special professional skills. He need not possess the highest expert skill; it is enough if he exercises the ordinary skill of an ordinary competent man exercising his particular art. So much is established by *Bolam* v. *Friern Hospital Management Committee* [1957] 1 WLR 582 which has been applied and approved time without number. "No matter what profession it may be, the common law does not impose on those whose practise it any liability for damage resulting from what in the result turn out to have been errors of judgment, unless the error was such as no reasonable well-informed and competent member of that profession could have made" (*Saif Ali* v. *Sydney Mitchell & Co* [1980] AC 198 220D, per Lord Diplock).

From these general statements it follows that a professional man should command the corpus of knowledge which forms part of the professional equipment of the ordinary member of his profession. He should not lag behind other ordinarily assiduous and intelligent members of his profession in knowledge of new advances, discoveries and developments in his field. He should have such awareness as an ordinarily competent practitioner would have of the deficiencies in his knowledge and the limitations on his skill. He should be alert to the hazards and risks inherent in any professional task he undertakes to the extent that other ordinarily competent members of his profession would be alert. He must bring to any professional task he undertakes no less expertise, skill and care than any other ordinarily competent members of his profession would bring, but need bring no more. The standard is that of the reasonable average. The law does not require of a professional man that he be a paragon, combining the qualities of polymath and prophet.

In deciding whether a professional man has fallen short of the standards observed by ordinarily skilled and competent members of his profession, it is the standards prevailing at the time of acts or omissions which provide the relevant yardstick. He is not ... to be judged by the wisdom of hindsight. This of course means that knowledge of an event which happened later should not be applied when judging acts and/or omissions which took place before that event ... ; ... it is necessary, if the defendant's conduct is to be fairly judged, that the making of [any] retrospective assessment should not of itself have the effect of magnifying the significance of the ... risk as it appeared or should reasonably have appeared to an ordinarily competent practical man with a job to do at the time.'

State-of-the-art defence

The policy of the courts in avoiding hindsight when judging negligence and applying the test of the professional knowledge of the time the design was carried out leads to what is known as the state-of-the-art defence. That is the designer applied the knowledge of the time.

This defence is open to challenge on a number of grounds; one of which is that there may be a continuing duty on a designer to check his design and to revise in the light of new knowledge. How far this extends to a duty to warn after completion is a subject beyond the scope of this book since the law is still developing and definite principles are hard to define.

Another possible challenge is that the practice of the profession at the time of the design was itself incorrect. This challenge has been successfully mounted in some medical negligence cases, most notably, *Sidaway* v. *Governors of Bethlehem Royal Hospital* (1985).

Innovation and special circumstances

There is then a third challenge to the state-of-the-art defence. It will not do to say that the work was both at and beyond the frontier of professional knowledge at that time. The law requires pioneers to be prudent.

This is how Lord Edmund-Davies explained the matter in the *IBA* v. *EMI* case:

'What is embraced by the duty to exercise reasonable care must always depend on the circumstances of each case. They may call for particular precautions (*Redhead* v. *Midland Railway Co* (1869) LR4 QB, 379 at 393). The graver the foreseeable consequences of failure to take care, the greater the necessity for special circumspection (*Paris* v. *Stepney Borough Council* [1951] AC367), and "Those who engage in operations inherently dangerous must take precautions which are not required of persons engaged in the ordinary routine of daily life" (*Glasgow Corporation* v. *Muir* [1943] AC 448, per Lord Macmillan at 456). The project may be alluring. But the risks of injury to those engaged in it, or to others or to both, may be so manifest and substantial, and their elimination may be so difficult to ensure with reasonable certainty that the only proper course is to abandon the project altogether. Learned counsel for BICC appeared to regard such a defeatist outcome as unthinkable. Yet circumstances can and have at times arisen in which it is plain commonsense, and any other decision foolhardy. The law requires even pioneers to be prudent.'

4.5 *Liability in tort*

It is not intended in this book to deal with liability in tort other than briefly. The key principles are as follows:

- there must be a duty of care;
- there must be breach of that duty;
- the breach must result in damage or injury.

The modern law of the tort of negligence derives from the famous snail in the ginger beer bottle case – *Donoghue* v. *Stevenson* (1932). In that case Lord Aitkin said:

'The rule that you are to love your neighbour becomes in law, you must not injure your neighbour; and the lawyer's question, Who is my neighbour? receives a restrictive reply. You must take reasonable care to avoid acts or omissions which you can reasonably foresee would be likely to injure your neighbour. Who, then, in law is my neighbour? The answer seems to be – persons who are so closely and directly affected by my act that I ought reasonably to have them in contemplation as being so affected when I am directing my mind to the acts or omissions which are called in question.'

For half a century the law developed adding new categories of negligence with physical damage and economic loss joining injury as grounds for action. But in the 1980s the courts began a reversal of the trend, culminating in the case of *Murphy* v. *Brentwood District Council* (1990). The position now is that for a physical damage claim in negligence the physical damage must be damage to an article or property other than the article or property to which the negligence itself applies – *D & F Estates* v. *The Church Commissioners* (1988). And, except in cases of special proximity such as *Hedley Byrne* v. *Heller* (1964), pure economic loss is no longer recoverable in negligence – *Murphy* v. *Brentwood* (1990).

Negligence

The meaning of negligence was eloquently stated in the case of *Bolam* v. *Friern Hospital Management Committee* (1957). Bolam, a patient in a mental hospital, was given electro-convulsive therapy without any restraint or sedation and as a result was injured. Mr Justice McNair said:

'I must tell you what in law we mean by "negligence". In the ordinary case which does not involve any special skill, negligence in law means a failure to do some act which a reasonable man in the circumstances would do, or the doing of some act which a reasonable man in the circumstances would not do; and if that failure or the doing of that act results in injury, then there is a cause of action. How do you test whether this act or failure is negligent? In an ordinary case it is generally said you judge it by the action of the man in the street. He is the ordinary man. In one case it has been said you judge it by the conduct of the man on the top of a Clapham omnibus. He is the ordinary man. But where you get a situation which involves the use of some special skill or competence then the test as to whether there has been negligence or not is not the test of the man on the top of a Clapham omnibus, because he has not got this special skill. The test is the standard of the ordinary skilled man exercising and professing to have that special skill. A man need not possess the highest expert skill; it is well established law that it is sufficient if he exercises the ordinary skill of an ordinary competent man exercising that particular art.'

4.6 *Design in MF/1 and MF/2*

When MF/1 was introduced in 1988 it included for the first time in the

IMechE/IEE Conditions provisions covering the contractor's obligations and responsibilities for design. Similar provisions are now included in MF/2.

Design obligations

Clause 13.1 of MF/1 (and MF/2) includes design in the list of the contractor's general obligations. And the specimen forms of agreement for both MF/1 and MF/2 also include design in the contractor's obligations.

Slightly oddly perhaps the specimen forms of tender do not expressly refer to design – the contractor merely offers to supply the plant and execute the works. Perhaps the definition of the 'Works' in the contracts is thought to be wide enough to include for design. Or perhaps 'supply' is deemed to include for design.

The question of whether the contractor's obligation is to design all or part of the works is not directly addressed. Unlike the IChemE model forms and some other design and construct forms which use the phrase 'complete the design' both MF/1 and MF/2 refer to design in clause 13.1 as though the contractor has the complete obligation.

But this is not consistent with clause 13.3 which refers to 'any detailed design provided by the Purchaser or the Engineer' nor with clause 5.6 which refers to work executed under prime cost items and (in the case of MF/1) under provisional sums.

Nevertheless the intention of both contracts is almost certainly that the contractor is obliged to design all the works subject to the above exceptions – and that his price should include for doing so. To avoid disputes on this, particularly with regard to work on site, the specification should ideally detail those works which the contractor is not obliged to design.

The alternative course of detailing the works the contractor is obliged to design runs the risk that items may be accidentally omitted and may have to be reinstated as variations.

Design responsibility

Clause 13.3 of MF/1 (and MF/2) is the principal clause on design responsibility.

Clause 13.3 – contractor's design

The Contractor shall be responsible for the detailed design of the Plant and of the Works in accordance with the requirements of the Specification. In so far as the Contractor is required by the Contract or is instructed by the Engineer to comply with any detailed design provided by the Purchaser or the Engineer the

Contractor shall be responsible for such design unless within a reasonable time after receipt thereof he shall have given notice to the Engineer disclaiming such responsibility.

Unless otherwise provided in the Contract the Contractor does not warrant that the Works as described in the Specification or the incorporation thereof within some larger project will satisfy the Purchaser's requirements.

The first sentence of this clause can be seen as relating to liability. Providing the contractor's design meets the requirements of the specification the contractor has fulfilled his obligations in respect of design.

In a limited sense this excludes fitness for purpose in that if the specification does not meet the purchaser's requirements that is not the contractor's problem or default. However, in so far that the specification is performance-related then liability for design is also performance-related – which amounts to much the same thing as fitness for purpose. And this, of course, ties in with the provisions in the contract for rejecting plant which does not pass specified tests.

So to the extent that plant which does not match the specification and does not pass specified tests can be rejected the contractor has a responsibility to design to a purpose – to meet the specification and to pass the tests. It is not open to the contractor to argue that providing he uses skill and care in his design he fulfils his contractual obligations whether or not the plant meets the specification or passes the tests.

That view might be challenged by quoting the phrase 'with due care and diligence' which precedes the word 'design' in clause 13.1 (contractor's general obligations). But that phrase it is suggested is intended to emphasize the contractor's obligations and not to give relief from them.

Assumed responsibility

The second sentence of clause 13.3 relates to designs provided by the purchaser or the engineer. The clause provides that the contractor is responsible for such designs unless he gives notice disclaiming responsibility within a reasonable time of receipt of the designs.

The intention of this is to achieve single point responsibility for the design if the contractor raises no objection. It is not an uncommon arrangement in the plant industry and similar provisions are found in various other standard forms with contractor's design. There are, however, a number of difficulties to consider.

Negligence in the design

The first of these is how far – by accepting responsibility for the purchaser's or the engineer's design – the contractor is liable for the consequences of any negligence in that design. As between the parties it seems to be intended that the contractor should be fully liable. This follows from the exclusion in the purchaser's risks in clause 45.1 of MF/1 of any design for which the contractor has not disclaimed responsibility.

However if there is injury or damage to third parties the position is not straightforward. By clause 43.4 of MF/1 the contractor is not liable for such injury or damage occurring whilst the contractor has responsibility for care of the works if the cause is due to the purchaser's or the engineer's negligence. However, by clause 43.5 of MF/1 the contractor is liable for and expressly indemnifies the purchaser against claims for such injury or damage occurring after responsibility for care of the works passes to the purchaser and which is due to design for which the contractor has not disclaimed responsibility.

The reason for the disparity is not obvious. But it would seem that until take-over the contractor is not liable for third party claims arising from negligence in the purchaser's or the engineer's design whether or not he has made a disclaimer of responsibility for the design. It is understood that the 1995 Revision 3 of MF/1 will amend clause 43.4 to bring it into line with clause 43.5.

Consequences of disclaiming responsibility

The second, and the more commonly encountered difficulty with clause 13.3, is the position which arises when the contractor disclaims responsibility for the purchaser's or the engineer's design.

This can leave the purchaser without any legal remedy in the event of failure unless the purchaser has obtained warranties in respect of the disclaimed designs. And it puts into question how the provisions for tests on completion and performance tests are intended to operate in the event of defective design for which the contractor is not responsible. Neither MF/1 nor MF/2 deal directly with this but in practice if the works cannot be satisfactorily completed and tested because of such defective design then usually the engineer will be obliged to order variations.

Meaning of design

Fundamental to these matters is what is meant by 'detailed design' in clause 13.3 and what is a 'reasonable time after receipt thereof' for the contractor to give notice of disclaimer.

Design is not a defined term in MF/1 or MF/2 and as mentioned earlier in this chapter its ordinary meaning is wide. It is certainly possible that within the meaning of clause 13.3 'detailed design' covers specified (or named) specialist items of plant to be incorporated into the works. Such a meaning is consistent with the provisions of clause 5.6 of MF/1 (and MF/2) discussed below. But it could also be argued that 'detailed design' has a very narrow meaning in the context of the contract as a whole if conflict is to be avoided with clause 16.2 (errors in drawings etc supplied by purchaser or engineer). See the comment on clause 16.2 later in this section.

Time for disclaimer

As to what is a reasonable time for disclaimer that, as with all such references to a reasonable time, is a matter of fact to be determined on the circumstances of each case. But clearly the complexity of the design and the timing of its issue relevant to the progress of the works could be taken into account.

What is less certain is whether the contractor can, promptly or shortly after being awarded the contract, give notice of disclaimer of all the design in the specification provided at the tender stage. The contractor may argue in support of this that until the contract is awarded clause 13.3 does not come into play and there is no obligation at tender stage to disclaim responsibility for the design.

Perhaps it would be better in a matter as important as this if clause 13.3 made it clear whether it applies to design provided with the tender or only to designs provided after award of the contract. And if it does apply to tender designs whether disclaimers should be made by the contractor with, or before, submission of his tender.

No requirement for reasons

Note that the requirement for reasonableness in clause 13.3 on disclaiming responsibility for design applies only to the timing of the notice. There is no requirement that the contractor should give reasons showing that the notice itself is reasonable. The contractor has absolute discretion on this.

The purchaser's requirements

The final part of clause 13.3 makes the point that the contractor, in accepting responsibility to design and construct the works to the specification, does not warrant that the works will satisfy the purchaser's requirements.

In so far that the works are part of a larger project this must obviously be the case. But where the works are a complete operating unit this may seem, from the purchaser's viewpoint, an unreasonable provision.

The objection, however, is easily overcome. It is up to the purchaser to specify the performance he requires and to leave the detail of design to the contractor. The contractor then does effectively warrant that the plant will be fit for the performance level specified.

Design responsibility under provisional sum and price cost items

Notwithstanding the statement in clause 13.3 of MF/1 (and MF/2) that the contractor shall be responsible for design an important exception is provided in clause 5.6 (of both forms).

Clause 5.6 (MF/1)

> The Contractor shall have no responsibility for work done or Plant supplied by any other person in pursuance of directions given by the Engineer under this Clause unless the Contractor shall have approved the person by whom such work is to be done or such Plant is to be supplied and the Plant, if any, to be supplied.

In MF/1 clause 5.6 applies to work ordered under provisional sums and prime cost items but in MF/2 it applies only to prime cost items.

In both forms, however, the potential consequences to the purchaser of the contractor having no responsibility for parts of the works are serious. See the comments above on disclaimed responsibility and in Chapter 18 on provisional sums and prime cost items.

Effect of clause 16.2

By clause 16.2, the text of which is given in section 4.7 which follows, the purchaser is responsible for errors, omissions or discrepancies in drawings or written information supplied by him or the engineer. The question is, how is this to be reconciled with clause 13.3 which makes the contractor responsible for detailed design so supplied unless he makes a disclaimer? It would seem to be necessary to distinguish between 'detailed design' and 'drawings or written information'. It is doubted if this will always be practicable.

Statutory consents

An important aspect of design and construct contracts is which party is responsible for obtaining statutory consents and the like relating to the works. In many cases, particularly where there are buildings to erect or other structures requiring planning consent this impinges directly on the design.

In MF/1 (and MF/2) the responsibility falls on the purchaser under clause 11.2 and this must be allowed for by the engineer and the purchaser in the preparation of the contract. It is too late after the contract has been awarded for the purchaser to find out there is a problem with statutory consents. By then that problem is the purchaser's problem and the contract provides for the contractor to be compensated for any delay.

4.7 Drawings in MF/1 and MF/2

Approval of contractor's drawings

Clause 15.1
This applies in both MF/1 and MF/2.

The Contractor shall submit to the Engineer for approval:

(a) within the times given in the Contract or in the Programme such drawings, samples, patterns, models or information (including calculations) as may be called for therein, and in the numbers therein required;
(b) during the progress of the Works within such reasonable times as the Engineer may require such drawings of the general arrangement and details of the Works as may be specified in the Contract or as the Engineer may reasonably require.

The Engineer shall signify his approval or disapproval thereof. If he fails to do so within the time given in the Contract or the Programme, or if no time limit is specified, within 30 days of receipt, they shall be deemed to have been approved.
Approved drawings, samples, patterns and models shall be signed or otherwise identified by the Engineer.
The Contractor shall supply additional copies of approved drawings in the form and numbers stated in the Contract.

The primary purpose of this clause is to enable the engineer to check that the contractor's design complies with the specification and that it has a compatible interface with other works.

The key question is, what are the consequences of the engineer's approval; whether given in writing or deemed to have been given?

It might appear from clause 16.1, which states that notwithstanding approval the contractor is responsible for any errors or omissions, that approval counts for little. But clause 16.1 is limited in its scope to errors and the like and it does not negate generally the consequences of approval.

A more significant provision is in clause 15.3 which states that approved drawings shall not be departed from except as provided in clause 27 (variations). This suggests that approval does confer on the 'drawings' – if not on the rest of the contractor's design – a contractual status equal or similar to that of contract drawings.

Consequences of disapproval of drawings

Clause 15.2

Any drawings, samples, patterns or models which the Engineer disapproves, shall be modified and re-submitted without delay.

This clause obliges the contractor to respond to disapproval. The contractor is not entitled to proceed with the works to his original drawings, etc on the proposition that he alone carries responsibility for design.

The role of the engineer in approving drawings is a contentious matter in any design and construct contract because of the way it cuts across the responsibilities of the contractor. But in contracts with performance specifications – as is often the case in plant contracts – the engineer's role is even more questionable.

Engineers should certainly be alert to the possibility that comments they make in disapproving the engineer's design and drawings may come in for close scrutiny as instructions under clause 13.3 or variations under clause 27.1.

As to the situation when there is an unresolvable dispute between the contractor and the engineer on whether designs and drawings are in compliance with the specification, the contractual position under MF/1 is as follows:

- under clause 2.4 the contractor must proceed with the works in accordance with the instructions and orders of the engineer;
- the contractor can give notice of dispute under clause 2.6;
- the dispute can be referred to arbitration under clause 52.1;
- performance of the works is to continue during arbitration unless the engineer orders suspension.

Approved drawings

Clause 15.3

Approved drawings shall not be departed from except as provided in Clause 27 (Variations).

See the comment above on clause 15.1.

Inspection of drawings

Clause 15.4

> The Engineer shall have the right at all reasonable times to inspect all drawings of any part of the Works.

It is not clear if this clause is intended to apply to drawings in the course of preparation or only to drawings which are ready for or have been submitted for approval.

As a general rule it would be unwise for the engineer to become involved in the contractor's design process as 'instructions' could inadvertently be given under clause 13.3.

Foundations, drawings, etc

Clause 15.5

> The Contractor shall provide within the times stated in the Contract or in the Programme, drawings showing how the Plant is to be affixed and any other information required for
>
> - preparing suitable foundations or other means of support,
> - providing suitable access for the Plant and any necessary equipment to the point on Site where the Plant is to be erected, and
> - making necessary connections to the Plant.

This clause relates principally to information required from the contractor to enable associated works to proceed. Not uncommonly projects comprise the supply and erection of plant under MF/1 and the simultaneous construction of civil engineering and building works under other contracts.

To protect the purchaser from delay claims on associated contracts it is not uncommon for the times for the supply of drawings, etc to be stipulated in the special conditions of plant contracts and for provision to be made for liquidated damages for late supply.

Operating and maintenance instructions

Clause 15.6

> Within the time or times stated in the Programme the Contractor shall supply operating and maintenance instructions and drawings of the Works as built.

These shall be in such detail as will enable the Purchaser to operate, maintain, dismantle, reassemble and adjust all parts of the Works.

Instructions and drawings shall be supplied in the form and numbers stated in the Contract.

The Works shall not be considered to be complete for the purposes of taking-over until such instructions and drawings have been supplied to the Purchaser.

This is an important clause and similar provisions are now being introduced into most design and construct contracts. The clause has obvious practical application but its observance is also closely related to increased awareness of health and safety responsibilities.

As quoted above clause 15.6 is a little odd in that it refers only to times stated 'in the Programme'. It is understood that in the 1995 Revision 3 of MF/1 the phrase 'in the Contract or in the Programme' will be used.

The final provision in the clause which in effect allows for deferment of taking-over until operating and maintenance instructions have been supplied, relies for its application on the diligence of the engineer. There is no provision for revoking a taking-over certificate mistakenly given before the relevant instructions have been supplied.

Purchaser's use of drawings, etc supplied by the contractor

Clause 15.7

Drawings and information supplied by the Contractor may be used by the Purchaser only for the purposes of completing, maintaining, adjusting and repairing the Works. No licence is granted to the Purchaser to copy or use drawings or information so supplied in order to make or have made spare parts for the Works. Drawings or information supplied by the Contractor shall not without the Contractor's consent be used, copied or communicated to a third party by the Engineer or the Purchaser otherwise than as strictly necessary for the purposes of the Contract.

It is reasonably clear from this clause what the purchaser may not do with drawings, etc supplied by the contractor.

What is less clear is when the purchaser's rights to use drawings and information accrue and whether they survive termination of the contract. The probable answers are that the rights accrue from submission of the drawings, etc and that they survive termination for contractor's default (see clause 49.1) but not termination for the purchaser's default.

Contractor's use of drawings, etc supplied by the purchaser or engineer

Clause 15.8

> Drawings and information supplied by the Purchaser and the Engineer to the Contractor for the purposes of the Tender and the Contract shall remain the property of the Purchaser. They shall not without the consent of the Purchaser be used, copied or communicated to a third party by the Contractor otherwise than as strictly necessary for the purposes of the Contract.

This clause allows the contractor to copy drawings, etc to sub-contractors and suppliers for the purposes of the contract.

Contractors should ensure that sub-contracts have back-to-back provisions as a safeguard against misuse at that level.

Manufacturing drawings, etc

Clause 15.9

> Notwithstanding any other provisions of the Contract the Contractor shall not be required to provide to the Purchaser or to the Engineer shop drawings nor any Contractor's confidential manufacturing drawings, designs, or know-how nor the confidential details of manufacturing practices, processes or operations.

Provisions of this type are common in all plant and process contracts but clearly there can be some contracts where it is entirely appropriate that the purchaser should have full knowledge and sight of manufacturing drawings, etc. In such cases the special conditions should specify what is required.

By clause 4.1 the special conditions prevail over the general conditions so clause 15.9 will then be of limited application. Alternatively clause 15.9 can be deleted.

Errors in drawings, etc supplied by the contractor

Clause 16.1

> Notwithstanding approval by the Engineer of drawings, samples, patterns, models or information submitted by the Contractor, the Contractor shall be responsible for any errors, omissions or discrepancies therein unless they are due to incorrect drawings, samples, patterns, models or information supplied by the Purchaser or the Engineer.
>
> The Contractor shall bear any Costs he may incur as a result of delay in providing such drawings, samples, patterns, models or information or as a

result of errors, omissions or discrepancies therein, for which the contractor is responsible.

The Contractor shall at his own expense carry out, or bear the reasonable Cost of, any alterations or remedial work necessary by such errors, omissions or discrepancies for which he is responsible and modify the drawings, samples, patterns, models or information accordingly.

The performance of his obligations under the Clause shall be in full satisfaction of the Contractor's liability under this Clause but shall not relieve him of his liability under Clause 34 (Delay).

This clause contains five distinct provisions as follows:

- The contractor is responsible for errors, etc in his drawings, etc unless the cause is incorrect information supplied by the purchaser or the engineer.
- The contractor is to bear the costs he incurs from errors for which he is responsible.
- The contractor is to bear the costs he incurs as a result of his delay in providing drawings, etc.
- The contractor shall bear the costs of remedial works arising from errors for which he is responsible.
- The contractor's liability for errors is limited to the costs referred to in the clause plus any damages for delay in completion.

The first three of these provisions are all expressions of ordinary legal liability. The fourth provision is again compatible with ordinary legal liability but it is not clear as to its scope. It obviously includes remedial work to the works of the contract but it is arguable as to whether it extends to paying for remedial works necessitated in other contracts.

The fifth provision which is in effect a limitation of liability is most obviously intended to prevent the purchaser from passing on claims for delay and disruption from other contractors – the purchaser's own claims for consequential loss are already excluded by clause 44.2. But whether it also extends to the direct cost of remedial works to other contracts as considered above is again arguable.

Errors in drawings, etc supplied by the purchaser or engineer

Clause 16.2

The Purchaser shall be responsible for errors, omissions or discrepancies in drawings and written information supplied by him or by the Engineer. The Purchaser shall at his own expense carry out any alterations or remedial work necessitated by such errors, omissions or discrepancies for which he is responsible or pay the Contractor the Cost incurred by the Contractor in carrying out in accordance with the Engineer's instructions any such alterations or remedial work so necessitated.

Comment on this clause has been made above and is also made in Chapter 5. The point is made there that should the purchaser elect to carry out alteration or remedial work himself the risks are considerable.

As to the contractor's remedies this clause entitles him to recover only 'costs'. The contractor has no other entitlements because of the limiting effects of clauses 44.2 and 44.4.

4.8 Patents in MF/1 and MF/2

MF/1 and MF/2 have similar provisions on patents. They deal with indemnities for infringements and effects of infringements on performance of the contract.

Indemnity against infringement

Clause 42.1 of MF/1

Clause 30.1 of MF/2
These clauses require the contractor to indemnify the purchaser against claims for patent infringements. But the indemnities do not cover:

- use of the plant other than for the purpose indicated or to be inferred from the contract;
- infringements due to use of the plant in association with plant not supplied by the contractor.

See also clause 42.3 of MF/1 (30.3 of MF/2) for further exceptions relating to the purchaser's design.

Conduct of proceedings

Clause 42.2 of MF/1

Clause 30.2 of MF/2
These clauses provide that:

- if a claim is brought against the purchaser;
- the contractor shall be notified and he may conduct negotiations and proceedings;

- the purchaser shall not make any prejudicial admissions;
- the contractor shall provide security for the purchaser's costs and liabilities;
- the purchaser shall afford the contractor all available assistance in contesting any claim.

Purchaser's indemnity against infringement

Clause 42.3 of MF/1

Clause 30.3 of MF/2
These clauses apply to patent infringements caused by designs or instructions given by the purchaser or the engineer. They reverse the position of the contractor and the purchaser in the two preceding clauses so that the purchaser indemnifies the contractor.

Infringement preventing performance

Clause 42.4 of MF/1

Clause 30.4 of MF/2
These are practical clauses of a type not found in many other design contracts. They cover the position where an injunction is served on one of the parties for breach of patent or other intellectual property rights.

If the party at fault is unable to obtain the removal of the injunction within 90 days then:

- if the infringement is the subject of the contractor's indemnity the purchaser is entitled to terminate the contract under clause 49 (contractor's default);
- if the infringement is the subject of the purchaser's indemnity the contractor is entitled to terminate the contract under clause 51 (purchaser's default).

4.9 Design, drawings and patents in MF/3

Design

MF/3 mentions design only in respect of patents. However it is implicit that in a contract for the supply of goods that the vendor will normally be responsible for design.

Both the specimen form of tender and the specimen form of agreement refer only to 'supply'.

Drawings

Clause 3.1 – vendor's drawings, etc

> Drawings, illustrations, descriptions, price lists, and catalogues issued by the Vendor shall not form part of the Contract unless incorporated therein by reference or otherwise.

This is to protect the vendor from claims of misrepresentation and the like and to ensure that the contractual requirements are specifically drafted.

Clause 3.2 – supply of particulars and drawings

> The Vendor shall within a reasonable time supply to the Purchaser:
>
> (a) the particulars and drawings (if any) called for in the Contract, and
> (b) such drawings (other than shop drawings) and other particulars of the goods as may be reasonably necessary for the purposes of installation and maintenance (including such dismantling and re-assembling as maintenance may involve).

MF/3 has nothing to say on the subject of errors in drawings and if claims should arise they would be settled on common law principles. The vendor would no doubt argue if a claim was brought against him by the purchaser for the recovery of costs caused by errors in drawings that the damage was too remote.

Patents

Clauses 12.1, 12.2 and 12.3 of MF/3 match the provisions in clauses 42.1, 42.2 and 42.3 of MF/1 in that each party indemnifies the other against patent infringements and is entitled to take over the conduct of proceedings.

MF/3 has no equivalent of clause 42.4 of MF/1 covering termination in the event of an injunction.

4.10 Design in G/90

G/90 follows the old MF 'A' and remains virtually silent on design obligations, responsibilities and liabilities.

Design obligations

There are no specimen forms of tender or agreement indicating the obligations of the contractor nor is there any clause in the contract defining the general obligations of the contractor. Broadly the position is that it is implied that the contractor has such design obligations as follow from the specification.

Design responsibility

There is nothing in G/90 comparable to clause 13.3 of MF/1 making the contractor responsible for all of the design. There are however provisions confirming the opposite. Thus clause 32(iii) deals with responsibility for work executed under provisional sums and prime cost items in the same way as clause 5.6 of MF/1.

Clause 32(iii)

The Contractor shall have no responsibility for work done or Plant supplied by any other person in pursuance of directions given by the Engineer under this clause unless the Contractor shall have approved the person by whom such work is to be done or such Plant is to be supplied and the Plant, if any, to be supplied.

As for the responsibility for obtaining statutory consents and the like this is another matter on which G/90 is silent.

Design liability

The nearest that G/90 comes to addressing the matter of design liability is in clause 30(vii).

Clause 30(vii)

Where the Specification sets out the Purchaser's specific design requirements for any part of the Works, compliance with such requirements for that part shall, subject to Sub-Clause (i) of this Clause, cause that part to be considered fit for the purpose for which it was intended. Nothing in this Clause shall affect the liability of the Contractor under Clause 21 (Liability of Accidents and Damage).

However on examination it can be seen that what this clause is saying is not that the works as designed are to be fit for their purpose but that provided the contractor complies with the purchaser's requirements the works are deemed to be fit for the purpose for which they are intended. This is similar to the position in MF/1.

4.11 Drawings in G/90

Submission and approval of drawings

Clause 4(i)

The Contractor shall submit to the Engineer for approval:

(a) within the time specified in the Specification or if no time is specified then a reasonable time such drawings, samples, patterns, and models as may be called for therein, and

(b) during the progress of the Works, and within such reasonable times as the Engineer may require, such drawings of the general arrangement and of details of the Works or any part thereof as the Engineer may reasonably require, provided that the Contractor shall not be under any obligation to supply copies of shop drawings.

Within a reasonable period after receiving such drawings, samples, patterns, and models the Engineer shall signify his approval or otherwise. If the Engineer shall not approve any drawing, sample, pattern or model thus submitted the same shall be forthwith modified to meet the reasonable requirements of the Engineer and shall be re-submitted. Copies of all drawings which require to be approved by the Engineer shall be provided in triplicate by the Contractor. Drawings, samples, patterns, and models when approved shall if required by either party be signed or identified by both parties and if required by the Contractor one copy shall be returned to him. All dimensions marked on drawings shall be considered correct although measurements by scale may differ therefrom. Detailed drawings shall take precedence where they differ from general arrangement drawings.

The provisions in this clause include those in clauses 15.1, 15.2 and 15.9 of MF/1 and more detailed provisions relating to the accuracy of drawings.

Approved drawings

Clause 4(ii)

Drawings, samples, patterns and models approved as above described shall not be departed from except as provided in Clause 19 (Variations and Omissions).

Again this is similar to MF/1 (see clause 15.3). It appears to confirm that approval of drawings confers contractual status.

Inspection of drawings

Clause 4(iii)

> The Engineer shall have the right at all reasonable times to inspect at the premises of the Contractor all drawings of any portion of the Works.

Again as MF/1. See clause 15.4 of MF/1 for comment.

Operating and maintenance information

Clause 4(iv)

> The Contractor shall, if so desired by the Purchaser, furnish to the Purchaser in writing at the commencement of the maintenance period, or at such earlier times as may be named in the Specification, such information, accompanied by drawings, as may be necessary to enable the Purchaser to operate, maintain, dismantle, reassemble, and adjust all parts of the Works.

This is not as positive a provision as in clause 15.6 of MF/1 which defers taking-over until the relevant operating and maintenance information has been supplied.

Purchaser's use of drawings etc.

Clause 4(v)

> Drawings submitted in pursuance of paragraphs (a) and (b) of Sub-Clause (i) of this clause shall not, without the consent of the Contractor, be used by the Purchaser except for purposes of the Contract, nor shall they without such consent be communicated to third parties save in so far as may be necessary for the proper execution of the Works.

This is the same as clause 15.7 of MF/1.

G/90 however does not contain the reciprocal of clause 15.8 of MF/1 restricting the contractor's use of the purchaser's drawings.

Mistakes in information

Clauses 5(i) and 5(ii)

> 5(i)   The Contractor shall be responsible for, and shall pay the extra cost, if any, occasioned by any discrepancies, errors, or omissions in the drawings and other particulars supplied by him, whether they have been approved by the Engineer or not, provided that such discrepancies, errors or omissions be not due to inaccurate information or particulars furnished in writing to the Contractor by the Purchaser or the Engineer.
>
> 5(ii) The Purchaser shall be responsible for, and shall pay the extra cost, if any, occasioned by any discrepancies, errors, or omissions in the drawings and information supplied in writing by the Purchaser or the Engineer.

These clauses differ from the provisions in clauses 16.1 and 16.2 of MF/1 in omitting any mention of limitation of liability.

4.12 Patents in G/90

Clauses 7(i), 7(ii) and 7(iii) of G/90 are virtually the same as clauses 42.1, 42.2 and 42.3 of MF/1. See the section on MF/1 for comment.

G/90 does not contain any equivalent of clause 42.4 of MF/1 covering termination in the event of injunctions.

Chapter 5

Obligations of the purchaser

5.1 Introduction

Defining the obligations of the parties is one of the key functions of any contract. But it is not always possible to identify from the text of the contract alone the full extent of those obligations. Some contracts place greater reliance on implied terms than others and some as a matter of drafting policy keep detail to the minimum. Consequently although standard forms of contract intended to serve much the same purpose can vary in length, from less than 5000 words to more than 50,000 words, the obligations of the parties do not necessarily increase tenfold in proportion.

Since the amount of detail is so variable, the length of text in a contract devoted to any particular obligation is rarely a good guide to the importance of that obligation. It is a feature of many standard forms that the lesser the obligation the more text it attracts. The question then is – how is the scale of importance of obligations in a contract indicated or to be ascertained?

One approach is to look to the termination provisions of the contract. They will indicate which obligations, when breached, entitle the innocent party to go beyond the normal remedy of damages. Another approach is to look to the way the parties have set out their intentions in broad terms in some documents such as the form of agreement.

Thus the form of agreement for MF/1, which is fairly typical, effectively states that the purchaser shall pay the contractor in consideration for the execution and completion of the works.

Payment

The primary obligation of the purchaser, therefore, is to pay the contractor. And in those contracts which do list the defaults of the purchaser which entitle the contractor to terminate this is invariably top of the list.

Moreover even if the obligation to pay is not formalized in a contract the law will recognize such an obligation. See the comments on *quantum meruit* in Chapter 17.

Possession of the site

In contracts where the works are to be performed on the land or in the
premises of the purchaser, the obligation of the purchaser to give possession of the site is clearly essential. In fact, this is so obvious that in
many contracts it is not stated and it is left to be implied that failure to
give possession is breach of a fundamental term of the contract entitling
the contractor to terminate.

Prevention

An obligation common to all contracts of whatever their nature is that one
party must not do anything to prevent the other party from performing
the contract.

Thus, Lord Blackburn in *Mackay* v. *Dick* (1881) said that:

'Where in a written contract it appears that both parties have agreed
that something should be done which cannot effectively be done unless
both concur in doing it, the construction of the contract is that each
agrees to do all that is necessary to be done on his part for the carrying
out of that thing though there may be no express words to that effect.'

And Lord Justice Vaughan Williams in *Barque Quilpue Ltd* v. *Bryant* (1904)
said:

'There is an implied contract by each party that he will not do anything
to prevent the other party from performing a contract or to delay him in
performing it. I agree that generally such a term is by law imported into
every contract.'

The principle of prevention is much relied on in claims by contractors. It
was confirmed in the case of *Merton LBC* v. *Stanley Hugh Leach Ltd* (1985)
where it was said that the employer should not hinder or prevent the
contractor from carrying out his obligations in accordance with the terms
of the contract and from executing the works in a regular and orderly
manner.

Express obligations of the purchaser

The express obligations of the purchaser in each of the model forms for
plant contracts are set out in schedules later in this chapter. They are taken
from clauses which use the phrase 'the Purchaser shall' since the word
'shall' is the usual indicator of an obligation.

The obligations of the purchaser fall into five broad categories:

- general obligations – in particular those relating to the execution of the works on the site;
- administrative obligations – those relating to the contract documents, the appointment of the engineer and confidentiality;
- testing and taking-over obligations – those detailing the facilities to be provided by the purchaser and the arrangements for testing;
- payment obligations – those covering terms and times for payments;
- damage and insurances – those detailing indemnity provisions and similar obligations.

All the above categories except general obligations are covered in other chapters of this book and accordingly only the purchaser's general obligations are examined in detail in this chapter.

Implied obligations of the purchaser

The subject of implied terms and obligations is discussed in Chapter 17 under claims. The point is made there that few contracts endeavour to detail in full the obligations of the parties.

As a general rule missing terms can be implied into contracts if it is necessary to give business efficacy; and that rule almost certainly applies to MF/3 and G/90. With MF/1 and MF/2, however, the exclusive remedy provisions (clause 44.4 of MF/1; clause 32.4 of MF/2) exclude obligations based on implied terms. The implications of this are, to say the least, profound.

In the context of this chapter the important matter is how far the purchaser is responsible for the performance of the engineer.

The purchaser and the engineer

Some contracts, such as the IChemE forms, leave no room for doubt that the purchaser is responsible for the performance of the engineer and that the purchaser accepts that responsibility as an obligation under the contract. Clauses are written into the contract to that effect.

But even in the absence of express provisions there is little doubt in most standard forms that the purchaser's responsibility can be implied.

The point came up in the *Merton* v. *Leach* case mentioned above. That case concerned the common complaint of late issue of instructions and drawings by the architect and this is how the judge expressed the relationships between the employer (building owner), the architect and the contractor. He said:

'It is to my mind clear that under the standard conditions the architect acts as the servant or agent of the building owner in supplying the contractor with the necessary drawings, instructions, levels and the like and in supervising the progress of the work and in ensuring that it is properly carried out. He will of course normally though not invariably have been responsible for the design of the work. There are very few occasions when a building owner himself is required to act directly without the intervention of the architect. To the extent that the architect performs these duties the building owner contracts with the contractor that the architect will perform them with reasonable diligence and with reasonable skill and care. The contract also confers on the architect discretionary powers which he must exercise with due regard to the interests of the contractor and the building owner. The building owner does not undertake that the architect will exercise his discretionary powers reasonably; he undertakes that although the architect may be engaged or employed by him he will leave him free to exercise his discretions fairly and without improper interference by him.'

The key sentence in this passage is:

'To the extent that the architect performs these duties the building owner contracts with the contractor that the architect will perform them with reasonable diligence and with reasonable skill and care.'

On this premise, therefore, leaving aside the question of whether MF/1 and MF/2 permit implied terms, it will normally be the case in plant and process contracts, as in construction contracts, that the purchaser is responsible for failures of performance of the engineer.

The purchaser as the engineer

In MF/3 there is no engineer and the purchaser is the contract administrator. In MF/1, MF/2 and G/90 there will normally be an engineer but each of the model forms provides for the purchaser to act as engineer in default of appointment of an engineer.

One effect of the purchaser so acting is that the purchaser takes on directly the obligations of the engineer. These are considered in Chapter 8 of this book and they make extensive lists.

5.2 Purchasers' general obligations

This section deals with those obligations of the purchaser related principally to the execution of the works on the site. In MF/1 they are grouped mainly into clauses 11 and 12 but in G/90 the corresponding clauses are more widely dispersed.

The general obligations of the purchaser in both MF/1 and G/90 are expressed in general terms which usually deserve amplification in the special conditions whether or not the particular clause refers to the special conditions.

It is to the benefit of both parties that obligations are clarified before the contract is made. But where they are not so clarified the purchaser is likely to be the loser where the contract apportions obligations generally in his direction.

Access and possession

Clause 11.1 – (MF/1)

The Purchaser shall give the Contractor access (but not exclusive access) to the Site on the date specified in the Contract. If no date is stated then access shall be given in reasonable time having regard to the Time for Completion. The Purchaser shall provide such roads and other means of access to the Site as may be stated in the Specification, subject to such limitations as to use as may be imposed.

Clause 16(i) – (G/90)

Subject to Sub-Clause (iv) of this clause, access to and possession of the Site shall be afforded to the Contractor by the Purchaser in reasonable time and, except in so far as the Specification may provide to the contrary, the Purchaser shall provide a road or railway suitable for the transport of all Plant and Contractor's Equipment necessary for the execution of the Works from a convenient point on a public throughfare suitable for such transport or on a railway available to the Contractor to the point on the Site where it is to be delivered or used.

Clause 16(iv) – (G/90)

The access to and possession of the Site referred to in Sub-Clause (i) hereof shall not be exclusive to the Contractor but only such as shall enable him to execute the Works.

It can be seen that in MF/1 the word 'access' is used with two meanings – possession of the site and entry to the site.

Note that in MF/1 the purchaser's obligations on roads and other means of access are only as stated in the specification whereas in G/90 (as in the IChemE forms) the purchaser has an obligation to provide suitable means of access unless excluded in the specification.

Neither access to nor possession of the site are mentioned in MF/2 or MF/3. Clearly there is no need to cover possession in supply and delivery contracts but the omission of provisions on access requires that the delivery terms cover the respective obligations of the parties on the matter.

Hours of work

Clause 19.1 – (MF/1)

Unless otherwise provided in the Contract the Purchaser shall give the Contractor facilities for carrying out the Works on the Site continuously during the normal working hours generally recognized in the district.

Clause 16(v) – (G/90)

The Purchaser shall give the Contractor facilities for carrying out the Works on the Site continuously during the normal working hours generally recognized in the district.

Wayleaves and consents

Clause 11.2 – (MF/1)

The Purchaser shall, within the times stated in the Programme or, if not so stated, before the time specified for delivery of any Plant to the Site, obtain all consents, wayleaves and approvals in connection with the regulations and by-laws of any local or other authority which shall be applicable to the Works on the Site.

This is an important provision, particularly in relation to planning consents and it leaves no doubt as to the scope of the purchaser's obligations. G/90 has no corresponding provision which is a serious omission. MF/2 and MF/3 have no need for corresponding provisions.

Permits, licences and duties

Clause 11.3 – (MF/1)

The Purchaser shall obtain all import permits or licences required for any part of the Plant or Works within the time stated in the Programme or, if not so stated, in reasonable time having regard to the time for delivery of the Plant and the Time for Completion. The Purchaser shall pay all customs and import duties arising upon the importation of Plant into the country in which the Plant is to be erected.

Clause 11.1 – (MF/2)

The Purchaser shall obtain all import permits or licences required for any part of the Plant within the time stated in the Programme or if not so stated in reasonable time having regard to the time for Delivery. The Purchaser shall pay or reimburse to the Contractor all customs and import duties arising upon the importation of Plant into the country in which the Plant is to be erected. In the event that the Purchaser shall fail to obtain such import permits or licences then the additional Cost reasonably incurred by the Contractor in consequence thereof shall be added to the Contract Price.

MF/1 and MF/2 are both model forms for home or overseas use. MF/3 and G/90 are intended for home use and have no corresponding provisions.

It is understood that the 1995 Revision 3 of MF/1 will clarify the last sentence of clause 11.3 so that the purchaser is required to 'pay or reimburse to the Contractor'. And it will also state the purchaser's obligation to pay the contractor's reasonable costs in the event of the purchaser failing to obtain the necessary permits and licences.

Foundations

Clause 11.4 – (MF/1)

Buildings, structures, foundations, approaches or work, equipment or materials to be provided by the Purchaser shall be provided within the time specified in the Contract or in the Programme, shall be of the quality specified and in a condition suitable for the efficient transport, reception, installation and maintenance of the Works.

Clause 16(ii) – (G/90)

If a building, structure, foundation, or approach is by the Contract to be provided by the Purchaser, such building, structure, foundation, or approach shall be in a condition suitable for the efficient transport, reception, installation, and maintenance of the Works.

In both MF/1 and G/90 the purchaser effectively warrants that conditions will be suitable for the execution of the works in the plant contract. The purchaser takes responsibility therefore for the quality of work undertaken by other contractors on which the plant contractor has to build and arguably the purchaser's responsibility extends to the accuracy of that work.

MF/2 and MF/3 have no place for corresponding provisions.

Purchaser's lifting equipment

Clause 11.5 – (MF/1)

> The Purchaser shall at the Contractor's request and expense operate for the purposes of the Works any suitable lifting equipment belonging to the Purchaser that may be available on the Site and of which details are given in the Special Conditions. The Purchaser shall during such operation retain control of and be responsible for the safe working of the lifting equipment.

Clause 9(iv) – (G/90)

> The Purchaser shall at the request of the Contractor and for the execution of the Works operate any suitable lifting equipment belonging to the Purchaser that may be available on the Site and the Contractor shall pay a reasonable sum therefor. The Purchaser shall during such operation retain control of and be responsible for the safe working of the lifting equipment but shall not be responsible for any negligence of the Contractor.

There is a significant difference between these clauses. In MF/1 the clause applies only to lifting equipment as detailed in the special conditions; in G/90 the clause applies to any suitable lifting equipment available on site.

The question of damage to the plant during lifting operations being undertaken by the purchaser is a contentious and complex issue. The question is which party takes the risk of damage occurring? In MF/1, negligence by the purchaser or his servants is a defined purchaser's risk and damage to the plant during lifting might well therefore fall outside the contractor's responsibility for care of the works. In G/90, however, negligence of the purchaser is not listed as an excepted risk. The contractor accepts responsibility, so far as care of the works is concerned, for all risks other than the excepted risks. This suggests that the purchaser is not liable under the contract. He may, however, have a liability in tort.

Utilities and power

Clause 11.6 – (MF/1)

> The Purchaser shall make available on the Site for use by the Contractor for the purposes of the Works such supplies of electricity, water, gas, air and other

services as may be specified in the Special Conditions. Such supplies shall be made available at the point(s) specified in the Contract on the terms mentioned in Sub-Clause 18.2 (Site Services).

Clause 18.2 – (MF/1)

The Contractor shall provide any apparatus necessary for the use of such supplies of electricity, water, gas and other services as are made available for the Contractor's use by the Purchaser under Sub-Clause 11.6 (Utilities and Power). The Contractor shall pay the Purchaser for such use at the rates specified in the Special Conditions or, if not so specified, at such reasonable rates as the Engineer shall determine.

Clause 9(iii) – (G/90)

The Contractor shall be entitled to use for the purposes of the Works such supplies of electricity, water, and gas as may be available therefor on the Site and shall pay to the Purchaser for such use such sum as may be reasonable in the circumstances, and shall at his own expense provide any apparatus necessary for such use.

Tests on site

Clause 11.7 – (MF/1)

Where the Contract provides for any tests on Site, the Purchaser shall unless otherwise stated in the Special Conditions, provide free of charge such fuel, electricity, skilled and unskilled labour, materials, stores, water, apparatus, instruments and feedstocks as may be requisite and as may reasonably be requested by the Contractor to enable the tests to be carried out effectively.

Clause 27(iv) – G/90

The Purchaser, except where otherwise specified, shall provide free of charge, subject to the provisions of Sub-Clause (v) of this clause, such labour, materials, electricity, fuel, water, stores, and apparatus, as may be requisite and as may be reasonably demanded to carry out such tests efficiently.

These obligations are both subject to the test of reasonableness. It is not therefore the case that the purchaser is obliged to provide all and everything required for tests.

The provision of labour is the most contentious item and here the word 'requisite' according to the *Concise Oxford Dictionary* is that which is required by circumstances or is necessary for success. The obligation of the purchaser, therefore, can be seen as extending to the provision of labour and other things which it is not practicable in the circumstances for the contractor to provide.

Assistance with laws and regulations

Clause 12.1 – (MF/1) and (MF/2)

> Where the Plant is to be erected outside the Contractor's country the Purchaser shall assist the Contractor to ascertain the nature and extent of and to comply with any laws, regulations, orders or by-laws of any local or national authority having the force of law in the country where the Plant is to be erected, or which may affect the Contractor in the performance of the obligations under the Contract. The Purchaser will if requested provide copies thereof and the Contractor shall reimburse the Cost thereof.

This provision is applicable only to works abroad and there is nothing comparable in MF/3 or G/90.

Note that the clause refers only to assisting the contractor. It does not make the purchaser directly responsible for ascertaining the appropriate laws and regulations.

Errors in drawings, etc

Clause 16.2 – (MF/1) and (MF/2)

> The Purchaser shall be responsible for errors, omissions or discrepancies in drawings and written information supplied by him or by the Engineer. The Purchaser shall at his own expense carry out any alterations or remedial work necessitated by such errors, omissions or discrepancies for which he is responsible or pay the Contractor the Cost incurred by the Contractor in carrying out in accordance with the Engineer's instructions any such alterations or remedial work so necessitated.

Clause 5(ii) – (G/90)

> The Purchaser shall be responsible for, and shall pay the extra cost, if any, occasioned by any discrepancies, errors, or omissions in the drawings and information supplied in writing by the Purchaser or the Engineer.

In both clauses the purchaser's obligation extends only to defects in drawings and information in writing. The contractor should clearly be careful therefore about acting on oral instructions and information.

For the implications of these clauses on the contractor's responsibility for design see Chapter 4.

Breach of the purchaser's obligations

Clause 11.8 – (MF/1)

> In the event that the Purchaser shall be in breach of any of his general obligations imposed by this Clause then the additional Cost reasonably incurred by the Contractor in consequence thereof shall be added to the Contract Price.

This clause, which is discussed further in Chapter 17 on Claims, when taken together with clause 44.4 (exclusive remedies) has the effect of limiting the purchaser's liability for breach of his obligations to those matters listed as obligations in clause 11 and to those other matters where express provision is made for additional payments.

5.3 Other obligations compared

Administrative obligations

For comments on the appointment of the engineer see Chapter 8. In G/90 the purchaser has greater freedom in replacement of the engineer than in MF/1. For comments on the contract documents see Chapter 3.

None of the model forms makes the execution of a formal agreement mandatory. In G/90 the purchaser always pays for the expenses of preparation but in MF/1 and MF/2 the contractor pays if he requests that the agreement be prepared. Under MF/1 and MF/2 the contractor pays for the performance bond whereas under G/90 the purchaser pays.

MF/1 and MF/2 deal with confidentiality but not MF/3 and G/90.

Testing and taking-over obligations

Testing and taking-over obligations are covered in detail in Chapters 9, 12 and 13. MF/1 has unique obligations in respect of performance testing in requiring the purchaser to shut down the works if required for adjustments.

Payment obligations

For detailed comment on payment obligations see Chapter 18. MF/1 and MF/2 detail fall-back provisions for terms of payment; MF/3 leaves the terms to be agreed; and G/90 outlines alternative schemes for interim payments. The stated times for payment are 30 days from the issue of certificates in MF/1 and MF/2; and 14 days in G/90.

Damage and insurances

For detailed comment see Chapter 15. MF/1 alone has provisions relating to damage to highways and bridges and the purchaser's obligations in respect of extraordinary traffic claims.

5.4 Express obligations of the purchaser in MF/1

clause 2.8	– not to appoint any person to replace or act with the engineer without the contractor's consent
clause 7.1	– to enter into a formal agreement if so required
clause 9.1	– to keep details of the contract confidential
clause 11.1	– to provide access to the site (but not exclusive access)
clause 11.2	– to obtain consents, approvals, etc
clause 11.3	– to obtain import permits and licences and to pay customs and import duties
clause 11.4	– to provide buildings, foundations, etc within time and of specified quality
clause 11.5	– to operate the purchaser's lifting equipment on the site
clause 11.6	– to make available power and water as a specified for use by the contractor
clause 11.7	– to provide power, materials and labour for tests
clause 12.1	– to assist the contractor to ascertain regulations, etc for works abroad
clause 16.2	– to accept responsibility for errors in information supplied by the engineer or the purchaser
clause 18.4	– to pay the contractor for the use of his equipment or provision of services
clause 19.1	– to give facilities for continuous work on site during normal working hours
clause 21.3	– to negotiate and settle and to indemnify the contractor in respect of extraordinary traffic claims
clause 35.4	– to permit adjustments to the works for performance tests and to shut down the works if so required
clause 35.7	– to evaluate with the contractor the results of performance tests
clause 36.3	– to inform the contractor of defects or damages (unless the engineer has already done so)
clause 36.8	– to bear the costs of searches for defects unless the contractor is responsible

clause 40.1   – to pay on certificates within 30 days or such other period as specified

clause 42.2   – to afford all available assistance to the contractor in contesting claims relating to patent rights, etc and not to make any prejudicial admissions thereon

clause 42.3   – to indemnify the contractor against patent infringements in respect of engineer/purchaser design

clause 43.4   – to indemnify the contractor against claims which are the inevitable consequence of the execution of the works

clause 43.6   – to indemnify the contractor against claims for accidents to workmen arising from purchaser's default

clause 43.7   – to afford all available assistance to the contractor for defending claims in respect of damage to persons and property and not to make any prejudicial admissions thereon

clause 46.4   – to pay sums due on termination for *force majeure*.

5.5 Express obligations of the purchaser in MF/2

clause 2.8   – not to appoint any person to replace or act with the engineer without the contractor's consent

clause 7.1   – to enter into a formal agreement if so required

clause 9.1   – to keep details of the contract confidential

clause 11.1   – to obtain import permits and licences and to pay customs and import duties

clause 12.1   – to assist the contractor to ascertain regulations, etc for works abroad

clause 16.2   – to accept responsibility for errors in information supplied by the engineer or the purchaser

clause 25.2   – to return to the contractor as required defective or damaged goods

clause 25.3   – to inform the contractor of defects or damages (unless the engineer has already done so)

clause 28.1   – to pay on certificates within 30 days or such other period as specified

clause 30.2   – to afford all available assistance to the contractor in contesting claims relating to patent rights, etc and not to make any prejudicial admissions thereon

clause 30.3   – to indemnify the contractor against patent infringements in respect of engineer/purchaser design.

5.6 Express obligations of the purchaser in MF/3

clause 2.1	– to furnish within a reasonable time all such further information as the vendor reasonably seeks to perform the contract; to pay the costs of unreasonable delay in failing to do so
clause 5.2	– to obtain replacements for rejected goods at reasonable prices and under competitive conditions where practicable
clause 6.1	– to be responsible for off-loading goods
clause 7.4	– to obtain replacements for undelivered goods at reasonable prices and under competitive conditions where practicable
clause 8.1	– to pay in respect of goods where the purchaser delays delivery for more than 14 days; to pay the cost of storing, protecting and insuring such goods
clause 9.3	– to accept that the liability of the vendor for goods lost or damaged in transit is only to replace the same
clause 10.1	– to pay on terms agreed or following delivery
clause 11.2	– to accept that the liability of the vendor for defects in the goods after delivery is only to repair or replace
clause 12.2	– to afford all reasonable assistance to the vendor for contesting claims on patent rights, etc; and not to make any prejudicial admissions thereon
clause 12.3	– to warrant the designs, etc supplied by the purchaser will not cause the vendor to infringe patent rights, etc
clause 13.3	– to decide forthwith whether or not to proceed with variations which the vendor maintains will prevent or prejudice the fulfilment of his contractual obligations
clause 17.1	– to pay value added tax.

5.7 Express obligations of the purchaser in G/90

clause 1	– to notify in writing to the contractor the name of the engineer
clause 3(i)	– to pay the expenses of procuring security for due performance
clause 3(iii)	– to pay the expenses of the formal contract agreement
clause 5(ii)	– to be responsible for and pay for errors, etc in information supplied to the contractor
clause 6(i)	– not to unreasonably withhold consent to assignment
clause 7(ii)	– to provide all reasonable assistance to the contractor for contesting claims on patent rights, etc; and not to make any prejudicial admissions thereon

clause 7(iii)	– to warrant that designs, etc given to the contractor will not infringe patents, etc
clause 9(iii)	– to make available on a charge basis supplies of water, gas, electricity already on site
clause 9(iv)	– to make available and operate on a charge basis, lifting equipment already on site
clause 14(iv)	– to provide free of charge, labour, materials, etc for site tests
clause 15(iv)(d)	– to assume responsibility for delayed plant after due notice
clause 16(i)	– to give access to and reasonable possession of the site; and to provide facilities for transport of plant and equipment from public access points
clause 16(ii)	– to provide in a condition suitable for the contractor's use any building, structures or foundations
clause 16(v)	– to give the contractor facilities on site for continuous working during normal working hours
clause 21(v)	– to afford the contractor all available assistance for defending claims for accidents and damages; and not to make any prejudicial admissions thereon
clause 23(ii)	– to bear the cost of repairing damage to the works caused by the purchaser up to the amounts of any stated policy excesses
clause 27(iv)	– to provide free of charge, labour, materials, etc for tests before completion
clause 28(iv)	– to pay the contractor in respect of interference with tests
clause 30(viii)	– to give the contractor right of access until the final certificate is issued
clause 34(i)	– to pay the contractor amounts certified by the engineer within 14 days
clause 35	– to pay value added tax.

Chapter 6

Obligations of the contractor

6.1 Introduction

In general terms the obligations of a contractor are to give performance as specified within the time allowed at the agreed price. And frequently in the form of tender and the form of agreement the contractor's obligations are not expressed in much greater detail.

To the extent that further detail is necessary it may either be expressed in the contract documents, in particular in the conditions of contract, the schedules and the specification, or it may be implied. But it is not good practice to confuse the detail of obligations with the scope of obligations. Detail expands on the scope; it rarely defines it.

Scope of the contractor's obligations

Scope by definition is intention or range. And clearly every contract should be precise as to the scope of the contractor's obligations.

This is best achieved by the use of key words rather than lengthy statements and typically these words include:

- design
- manufacturer
- test
- deliver
- supply
- erect
- construct
- complete
- remedy defects.

The words are not necessarily mutually exclusive – thus obligations expressed as 'design' and 'construct' might be deemed to include other obligations in the list. But for the avoidance of doubt the contract should contain a statement of obligations in the tender, the acceptance or the agreement which leaves no doubt as to what is intended. So while it may be adequate to state 'supply and erect' with the assumption that manufacture and delivery are automatically included it would be unwise to assume that 'construct and complete' implies any obligations to design. And ideally the combination of words used in the form of agreement should match the combination used in the form of tender. Surprisingly even with standard forms (including MF/1 and MF/2) this is not always the case.

Detail of the contractor's obligations

Detail in the conditions of contract on the contractor's obligations serves two purposes. It regulates how the contractor is to fulfil his several obligations and it sets out the subsidiary obligations which the contractor agrees to perform. These cover such matters as:

- bonds, guarantees and warranties
- serving of notices
- execution of the works
- testing and taking-over
- indemnities and insurances
- correction of defects
- applications for payments.

The amount of such detail varies from contract to contract. Thus in MF/1 over 80 express obligations/responsibilities can be identified (see section 6.3 which follows) but in MF/3 there is only a quarter of this number – a situation which follows naturally from the scope of the contract.

Obligations/responsibilities

The point is made in Chapter 4 on design that contracts sometimes take the words 'obligations' and 'responsibilities' to be synonymous. But there is a difference. An obligation is a burden to be undertaken; a responsibility is a burden to be carried. The contractor has an obligation to construct the works; he has responsibility for damage to the works. The contractor may have responsibility for design undertaken by others; but he has no obligation to do that design.

Because the model forms for plant contracts do use the words 'obligations' and 'responsibilities' with a certain amount of flexibility the schedules later in this chapter deal with both.

6.2 Obligations of the contractor in MF/1

If the specimen forms of tender and agreement are used these provide the starting point in registering the contractor's obligations.

Form of tender

The specimen form of tender for MF/1 goes further than many standard forms of tender in the detail of the obligations which the contractor offers to fulfil for the stated price. Thus the contractor, (or tenderer as he is at the time) expressly offers:

- to supply the plant and materials and to execute and carry out the works;
- to perform and observe the provisions of the conditions of contract, the specification, the schedules and the drawings;
- to execute a formal agreement if required to do so;
- to furnish a performance guarantee if required to do so;
- to obtain all the stipulated insurances;
- to do extra work ordered by the engineer.

It may be seen as something of an omission that there is no mention of design. This could be a point of significance if there is no intention to execute a formal agreement which does include design (as in the specimen form for MF/1).

The reference in the form of tender to extra work is unusual but welcome. It does emphasize a point so often overlooked – that the contractor is prepared to bind himself to doing more than just the specified work – he will take on variations. And the legal significance of this as discussed in Chapter 16 on variations is that any claims for breach of contract based on the issue of variations start from a very poor position.

Form of agreement

The specimen form of agreement for MF/1 states that the contractor agrees to 'design, manufacture, deliver to Site, erect, test and complete the Works and to remedy defects'. The form also states that the purchaser has accepted a tender from the contractor for these things.

It is not clear why there is such a disparity in the wording of the contractor's obligations between the specimen forms of agreement and tender – with the important matter of design in particular so conspicuously different. Some users of the forms change the wording to bring the express obligations into line.

Obligations in the general conditions

A schedule of express obligations and responsibilities of the contractor under MF/1 follows in the next section. Having regard to the extent and

Where the contractor has sole occupation of the site the application of this clause is relatively straightforward. But where there are other contractors on the site the boundaries of responsibility are not always clearcut.

The problems then arise – which contractor is to be made responsible for a particular function (e.g. the provision of temporary roadways) and does the contractor have a claim for additional payment if he takes on some responsibilities for other contractors.

Ideally the specification should indicate how functions are to be allocated between various contractors – and as to claims see clause 18.4 (opportunities for other contractors).

Site services

Clause 18.2 – site services

> The Contractor shall provide any apparatus necessary for the use of such supplies of electricity, water, gas and other services as are made available for the Contractor's use by the Purchaser under Sub-Clause 11.6 (Utilities and Power). The Contractor shall pay the Purchaser for such use at the rates specified in the Special Conditions or, if not so specified, at such reasonable rates as the Engineer shall determine.

By clause 11.6 the purchaser is obliged to make available to the contractor supplies of electricity, etc on the site. This clause confirms:

- the contractor is to provide the apparatus for the use of such supplies;
- the contractor is to pay for the use of such supplies.

Ideally the rates for supplies should be specified in the special conditions.

Clearance of site

Clause 18 – clearance of site

> From time to time during the progress of the Works the Contractor shall clear away and remove from the Site all surplus materials and rubbish and, on completion, all Contractor's Equipment. The Contractor shall at all times leave the Site and the Works clean and in a safe and workmanlike condition to the Engineer's reasonable satisfaction.

This clause confirms that it is not sufficient for the contractor merely to clear up on completion.

As with some other obligations of the contractor it might appear that there are inadequate remedies for default but again having regard to health and safety considerations the engineer has considerable powers under the contract (including suspension) to force the contractor to comply.

Hours of working

Clause 19.1 – hours of working

> Unless otherwise provided in the Contract the Purchaser shall give the Contractor facilities for carrying out the Works on the Site continuously during the normal working hours generally recognized in the district. The Engineer may, after consulting with the Contractor, direct that work shall be done at other times if it shall be practicable in the circumstances for work to be so done. The extra Cost of work so done shall be added to the Contract Price unless such work has, by the default of the Contractor, become necessary to ensure the completion of the Works within the Time for Completion.

This clause contains three distinct provisions:

- the purchaser is obliged to give the contractor facilities for working during normal working hours (unless otherwise provided);
- the engineer may direct that work shall be done at other times;
- the cost of working outside normal hours to the direction of the engineer can be recovered by the contractor providing it is not necessary to overcome his own delay.

The contractor's remedy for breach of the first of these is probably to be found in clause 11.8 (breach of the purchaser's general obligations). There is no other obvious remedy since the final provision of the clause appears to apply only to work outside normal hours.

If it is not intended that the contractor should work during normal working hours this should be specified in the special conditions.

The requirement in the second provision that the engineer should 'consult' with the contractor before directing work to be done outside normal working hours is open to various interpretations. It may be that the contractor is not in breach if he declines to comply with the engineer's directions if he has objected on the grounds that the proposed working hours are not practicable. Alternatively the consultation may be related to the estimates of costs.

The point in the third provision relating to default of the contractor is, it is suggested, applicable only where clause 14.6 (rate of progress) has been used by the engineer to require the contractor to take steps to remedy delay.

Clause 19.2 – no night or rest day working

> No work shall be carried out on Site during the night or on locally recognized days of rest without the consent of the Engineer or the Engineer's Representative unless the work is unavoidable or necessary for the protection of life or property or for the safety of the Works, in which case the Contractor shall immediately advise the Engineer or the Engineer's Representative. The Engineer or the Engineer's Representative shall not withhold any such consent if work at night or on rest days is considered by the Contractor to be necessary to meet the Time for Completion.

For contracts which expressly require night working and the like this clause has no application. Perhaps it should then be deleted.

With regard to the contractor's apparent right to work at night and on rest days – as indicated in the last sentence of the clause – this raises the question of conflict with the purchaser's obligation, as stated in clause 19.1, to give facilities for working during normal working hours. The answer, it is suggested, is that the contractor only has a right in so far that the purchaser is willing to make facilities available – in which case the engineer cannot withhold his consent. It is doubted if a breach of contract on the part of the purchaser can be established by arguing that the purchaser has an obligation to make night time facilities available and that failure to do so entitles the contractor to extra cost and/or extension of time.

Safety

Clause 20.1 – safety

> The Contractor shall be responsible for the adequacy, stability and safety of his operations on Site and shall comply with the Purchaser's safety regulations applicable at the Site unless specifically authorized by the Engineer to depart therefrom in any particular circumstances.

The contractor's first priority is, of course, to comply with all statutory requirements on safety. It is no defence for the contractor if he is in breach of these to say that he has a contractual requirement to comply with the purchaser's requirements or the directions of the engineer.

As to the apparently unambiguous, unqualified and sensible statement that the contractor is responsible for the adequacy, stability and safety of his operations note should be taken of a decision in the case of *Humber Oil Trustees Ltd* v. *Harbour and General Works Ltd* (1991).

In that case under the ICE Fifth edition, the Court of Appeal had to consider whether the contractor's obligation under clause 8(2) affected his rights of claim under clause 12 for unforeseen conditions. The contract was for the construction of mooring dolphins and the contractor selected and used a jack-up barge equipped with a fixed crane. As the crane was lifting and skewing a concrete soffit member it became unstable and collapsed. The contractor submitted a claim under clause 12 that he had encountered physical conditions which could not have been foreseen. The employer disputed this claim and argued that even if the contractor did encounter such conditions he had no claim under clause 12 by virtue of the provisions of clause 8(2). The clause stated that 'The Contractor shall take full responsibility for the adequacy, stability and safety of all site operations and methods of construction'.

The Court of Appeal rejected the employer's argument that as clause 8(2) was unqualified there was no room for the operation of clause 12. Lord Justice Nourse said:

> 'I cannot construe clause 8(2) as applying to a case where the inadequacy or instability is brought about by the contractor's having encountered physical conditions within clause 12(1). That I think was the instinctive view of the arbitrator, who dealt with this question very briefly. Like the judge, I have not found it an easy question, but like him and my Lord I would on balance decide it in favour of the contractors.'

The decision in the *Humber Oil Terminals* case has caused some concern. It cuts across the general principle that the contractor should be free to select his own methods of working but in doing so he must take responsibility for them. It means that the cost to the purchaser of the contract price will be dependent on how susceptible the contractor's chosen methods of working are to unforeseen conditions.

Setting out

Clause 22.1 – setting out

The Contractor shall accurately set out the Works in relation to original points, lines and levels of reference given by the Engineer in writing and provide all necessary instruments, appliances and labour therefor.

If, at any time during the execution of the Works, any error appears in the positions, levels, dimensions or alignment of the Works, the Contractor shall rectify the error.

The Contractor shall bear the Cost of rectifying the error, unless the error results from incorrect information supplied in writing by the Purchaser, the Engineer or the Engineer's Representative, or from default by another contractor, in which case the Cost incurred by the Contractor shall be added to the Contract Price.

The Contractor shall indemnify, protect and preserve bench marks, sight rails, pegs and other things used in setting out the Works.

There is repeated emphasis in this clause to information given to the contractor 'in writing'. The contractor should be aware that acting on information otherwise received is at his own risk.

To this there is one stated exception where the setting out error results from 'default by another contractor'. Perhaps this means that the contractor is not responsible for errors in following the work of other contractors. If this is correct, then in effect the purchaser warrants that the work of other contractors will be to the correct lines and levels.

Contractors frequently argue that there should be another exception – setting out checked and approved by the engineer or his representative but subsequently found to be in error. On this the clause is silent but the contractor will not find any grounds in the contract to support a claim for recovery of costs. There is no basic principle that approval by the engineer relieves the contractor of his responsibilities and many standard forms, although not MF/1, expressly state the opposite.

As to the final provision in clause 22.1 regarding the protection of setting out the case for such an obligation is sound but perhaps it would be more effective an obligation if there was a stated remedy for breach enabling the purchaser to recover wasted costs.

The word 'indemnify' in this provision is believed to be an error and it should probably be 'diligently'. It is understood the 1995 Revision 3 of MF/1 will delete the word 'indemnify'.

Vesting of plant and contractor's equipment

Clause 37.1 – ownership of plant

Plant to be supplied pursuant to the Contract shall become the property of the Purchaser at whichever is the earlier of the following times:

(a) when Plant is delivered pursuant to the Contract
(b) when the Contractor becomes entitled to have the Contract Value of the Plant in question included in an interim certificate of payment.

The reference to 'Contract Value' here is a little odd since there is no entitlement normally to have such a value included in an interim certificate of payment. This will probably be changed to 'value' in the 1993 Revision 3 to MF/1.

More generally in all contractual provisions relating to ownership of plant, materials, equipment etc it has to be recognized that the provisions are valid only between the parties to the contract. They do not affect the rights of third parties.

A well publicized building case under JCT Conditions which illustrated the point was *Dawber Williamson* v. *Humberside County Council* (1979).

A roofing sub-contractor brought slates on to site for which the contractor was paid by the employer. Before the slates were fixed the contractor went into liquidation. The sub-contractor attempted to retrieve the slates but the employer prevented this and claimed ownership. The sub-contractor then successfully sued the employer for their value. In considering the clause in the main contract which said that when the value of any goods has been included in any certificate under which the contractor has received payment, the goods shall be the employer's property, Mr Justice Mais said:

> 'In my judgment, this presupposes there is privity between the defendants and the sub-contractor, which there is not in the present instance, or the main contractor has good title to the material and goods. If the title has passed to the main contractor from the sub-contractor, then this clause has force.'

It should perhaps be added here that the law treats differently goods acquired under a sale of goods from goods acquired under a contract for work and materials (supply and fix). Goods acquired under a sale come within the scope of the Sale of Gods Act 1979. Under section 17 of that Act property is transferred at such time as the parties intend it to be transferred and regard is to be made, in ascertaining such intention, to the terms of the contract, the conduct of the parties and the circumstances of the case.

To avoid title passing before payment is made, many contracts for sale of goods often have clauses granting possession and even use to the buyer but retaining ownership with the seller until payment. These are sometimes called Romalpa clauses after the case of *Aluminium Industries* v. *Romalpa Aluminium Ltd* (1976).

Perhaps the only thing that can be said with certainty about ownership is that title passes to the purchaser when the plant is fixed to his property.

Clause 37.2 – marking of plant

> Where, prior to delivery, the property in Plant passes to the Purchaser, the Contractor shall, so far as is practicable, set the Plant aside and mark it as the Purchaser's property in a manner reasonably required by the Engineer.
>
> Until the Plant has been so set aside and marked the Engineer shall be entitled to withhold any interim certificate of payment to which the Contractor might otherwise be entitled.
>
> The Contractor shall permit the Engineer at any time upon reasonable notice to inspect any Plant which has become the property of the Purchaser and shall grant the Engineer or procure the grant of access to the Contractor's premises for such purposes or any other premises where such Plant may be located.
>
> All such Plant shall be in the care and possession of the Contractor solely for

the purposes of the Contract and shall not be within the ownership or disposition of the Contractor.

No interim certificate of payment issued by the Engineer shall prejudice his right to reject Plant which is not in accordance with the Contract. Upon any such rejection the property in the rejected Plant shall immediately revert to the Contractor.

Payment for materials and plant off-site is always a risky business because receivers or liquidators are not easily moved to recognize the rights of purchasers to items not fixed to their property.

That is not to say that the provisions of clause 37.2 should not be observed. It simply emphasizes that they should be observed meticulously. The purchaser will need all the strength he can muster to claim ownership if the contractor becomes bankrupt or goes into liquidation.

6.3 Express obligations and responsibilities of the contractor in MF/1

clause 2.5	– to request without delay that the engineer confirms instructions in writing
clause 3.2	– to be responsible for the acts, defaults and neglects of subcontractors
clause 5.1	– to inspect the site and to satisfy himself on all circumstances
clause 5.2	– to be responsible for interpretation of site data
clause 5.3	– to be responsible for incorrect information other than that provided in writing by the purchaser or the engineer
clause 5.7	– to inform the engineer of unexpected site conditions
clause 7.1	– to enter into a formal contract agreement if so required by the purchaser
clause 8.1	– to provide a performance bond if so required by the purchaser
clause 9.1	– to treat details of the contract as confidential
clause 13.1	– to design, manufacture, deliver, erect and test the plant, execute the works and carry out the tests on completion all within the time for completion
	– to made good defects
	– to provide specialist advice on performance testing
	– to provide all labour, supervision and contractor's equipment required for the execution of the works
clause 13.2	– to execute and manufacture the works in accordance with the specification
	– to carry out work on site in accordance with the reasonable directions of the engineer

clause 13.3 – to be responsible for detailed design
– to give notice of any disclaimer of responsibility for the purchaser's/engineer's design within a reasonable time of receipt
clause 14.1 – to submit a programme within 30 days of the letter of acceptance
clause 14.5 – to revise the programme if so required by the engineer
clause 14.6 – to expedite progress if so notified by the engineer
clause 15.1 – to submit drawings etc for approval and to supply additional copies of approved drawings
clause 15.2 – to modify and resubmit for approval drawings, etc which the engineer disapproves
clause 15.5 – to provide drawings, etc showing how the plant is to be fixed and other relevant information
clause 15.6 – to supply operating and maintenance instructions prior to taking-over
clause 16.1 – to be responsible for errors and omissions in drawings, etc unless they are due to information supplied by the purchaser or the engineer
– to carry out remedial work necessitated by drawings, etc for which the contractor is responsible and to modify drawings, etc accordingly
clause 17.1 – to employ one or more competent representatives to superintend the works on the site
clause 17.2 – to remove from site persons who in the opinion of the engineer are incompetent or negligent
clause 17.3 – to provide such returns of staff and labour as the engineer may require
clause 18.1 – to be responsible for fencing, watching, lighting, etc
clause 18.2 – to provide apparatus necessary for the use of site services made available by the purchaser and to pay reasonable rates for the use of such services
clause 18.3 – to clear the site from time to time and on completion
clause 18.4 – to afford reasonable opportunities for other contractors to carry out their works
clause 19.1 – to carry out work outside normal working hours if so directed by the engineer
clause 20.1 – to be responsible for the adequacy, stability and safety of site operations and to comply with the purchaser's safety regulations
clause 21.1 – to use reasonable means to avoid damage to highways or bridges
clause 21.2 – to give notice of special loads to the engineer and to propose and carry out protecting and strengthening measures
clause 21.3 – to report extraordinary traffic claims to the engineer

clause 22.1 – to accurately set out the works and provide all necessary instruments
    – to rectify errors in setting-out
    – to bear the cost of rectifying errors unless caused by information supplied by the purchaser, the engineer, or some other contractor
    – to protect and preserve bench marks

clause 23.1 – to obtain permission for the engineer to inspect, examine and test on sub-contractor's/supplier's premises

clause 23.2 – to agree dates for testing with the engineer and to supply certified copies of test results if the engineer does not attend

clause 23.5 – to repair or replace rejected plant and to resubmit for testing

clause 24.1 – to apply for permission to deliver plant to site and to be responsible for unloading all plant and contractor's equipment

clause 25.1 – to store, preserve, protect and insure the plant/works during periods of suspension
    – to maintain staff, labour and equipment on or near the site during periods of suspension

clause 25.5 – to make good deterioration, defects or loss occurring during suspension

clause 26.1 – to make good defects specified by the engineer before taking-over

clause 27.2 – to carry out variations instructed by the engineer

clause 27.4 – to keep contemporary records of the cost of variations

clause 27.5 – to notify the engineer if any variation is likely to prevent or prejudice the contractor from fulfilling his obligations

clause 27.6 – to immediately proceed with variations on receipt of the engineer's instructions

clause 28.1 – to give the engineer 21 days notice of readiness for tests on completion

clause 28.2 – to forward certified copies of test results if the engineer fails to attend tests on completion

clause 28.3 – to make delayed tests as instructed to do so

clause 28.4 – to carry out repeat tests if any part of the works fails to pass tests on completion

clause 28.5 – to take whatever steps are necessary to enable the works to pass tests on completion

clause 29.4 – to complete outstanding work within the time stated in the taking over certificate

clause 31.2 – to carry out tests on completion during the defects liability period when there has been interference with tests beforehand clause 32.1 – to complete the works and pass the tests on completion within the time for completion

clause 33.1 – to give notice of any claim for extension of time as soon as reasonably practicable after the delaying event

clause 33.3 – to consult with the engineer to mitigate delay and to comply with the engineer's instructions thereon

clause 34.1 – to pay damages for delay in completion

clause 35.2 – to assist the purchaser to carry out performance tests

clause 35.4 – to carry out adjustments and modifications to the plant at all reasonable speed to permit the repetition of performance tests

clause 35.7 – to evaluate with the purchaser the results of performance tests

clause 35.8 – to pay liquidated damages for failure to achieve guaranteed performance

clause 36.2 – to be responsible for making good defects or damage appearing or occurring during the defects liability period

clause 36.7 – to carry out further tests on completion as instructed subsequent to repairs or replacements

clause 36.8 – to search for defects if so required by the engineer

clause 36.10 – to make good defects appearing within three years of taking-over providing the defects were caused by 'gross misconduct'

clause 37.2 – to set aside and mark plant in which property has passed to the purchaser and to permit the engineer to inspect the same

clause 38.1 – to provide details within 30 days of the letter of acceptance of equipment to be used on site

clause 38.3 – to be liable for loss or damage to contractor's equipment

clause 38.4 – to be responsible for maintaining equipment in safe working order

clause 39.9 – to make application for the final payment certificate forthwith after completing defects liability obligations

clause 41.1 – to give notice of claims within 30 days of the circumstances arising and to submit full particulars of amounts claimed

clause 42.1 – to indemnify the purchaser against infringement of patents, etc

clause 43.1 – to be responsible for care of the works until taking-over

clause 43.2 – to make good loss or damage to the works prior to taking-over

clause 43.3 – to make good damage to the works caused by the purchaser's risks if so required

clause 43.4 – to indemnify the purchaser against third party claims prior to taking-over clause 43.5 – to indemnify the purchaser against third party claims after taking-over to the extent the claim arises from negligence of the contractor or his sub-contractors

clause 43.6 – to indemnify the purchaser against claims from workmen

clause 47.1 – to insure the works against loss or damage until 14 days after the issue of the taking-over certificate

clause 47.2 – to extend insurance cover as far as possible to the defects liability period

clause 47.3 – to apply insurance monies towards repair and replacement of the works

clause 47.4 – to insure against third party risks

clause 47.5 – to insure in respect of workmen

clause 47.6 – to effect insurances with an approved insurer and on approved terms

clause 48.2 – to procure the waiver of subrogation rights in respect of insurances in joint names

clause 51.2 – to remove contractor's equipment from site after giving notice of termination for purchaser's default.

6.4 Obligations of the contractor in MF/2

MF/2 is described in its title as a model form for the supply of plant and accordingly the obligations of the contractor are considerably less than those in MF/1 which includes erection. This is apparent from the schedule of express obligations and responsibilities which follows in the next section. It is also apparent from the specimen forms of tender and agreement included in MF/2.

Form of tender

In the specimen form of tender the contractor offers to:

- supply the plant
- observe the provisions of the contract
- execute a formal agreement if so required
- furnish a performance bond if so required
- obtain such insurances are stipulated
- do extra work ordered by the engineer.

As with the form of tender for MF/1 there is no express mention of the contractor's design obligation.

Forms of agreement

The contractor's obligations as stated in the form of agreement are to:

- design
- manufacture
- test
- deliver
- remedy defects.

As with MF/1, it is also stated that the purchaser has accepted a tender by the contractor to do these things. But since in the form of tender the contractor has offered only to 'supply' this is stretching the meaning of the word 'supply' to the limit.

Obligations in the general conditions

MF/2, like MF/1, excludes implied obligations. Clause 32.4 (exclusive remedies) indicates this by stating that the obligations in the conditions of contract are exhaustive of the parties' obligations.

The express obligations and responsibilities of the contractor are scheduled in the next section of this book. Comment on these is given in other chapters except for the matters listed below:

- clauses 5.1 to 5.3   – information/sufficiency of tender
- clauses 13.1 and 13.2 – general obligations
- clauses 26.1 and 26.2 – vesting of plant.

Information/sufficiency of tender

Clause 5.1 – contractor to inform himself fully

> The Contractor shall be deemed to have satisfied himself as far as can reason-ably be done as to all circumstances affecting the Contract (including any safety regulations applicable to the Plant in the country where it is to be erected), and to have examined the Conditions and Specification, with such drawings, schedules, plans and information as may be annexed thereto or referred to therein.

This differs from the corresponding clause in MF/1 only in that it excludes inspection of the site and substitutes safety regulations applic-able to a country for safety regulations of the purchaser.

Clause 5.2 – site data

> The Tender shall be deemed to have been based on such data on climatic, hydrological, soil and general conditions of the place where the Plant is to be erected and for the operation of the Plant as the Purchaser or the Engineer has made available in writing to the Contractor for the purposes of the Tender. The Contractor shall be responsible for his own interpretation of such data.

The same as clause 5.2 in MF/1 with 'the place where the Plant is to be erected' substituted for 'the Site'.

Clause 5.3 – incorrect information

> The Contractor shall be responsible for any misunderstanding or incorrect information however obtained except information provided in writing by the Purchaser or the Engineer.

Identical to clause 5.3 of MF/1.

General obligations

Clause 13.1 – general

> The Contractor shall, subject to the provisions of the Contract, with due care and diligence, design, manufacture, test and deliver the Plant within the Time for Delivery. The Contractor shall also make good defects in the Plant during the Defects Liability Period.

This clause is much shorter than the corresponding clause 13.1 in MF/1 because it is able to omit references to tests on completion, performance tests and resources for site operations.

Clause 13.2 – manner of execution

> The Plant shall be manufactured and executed in the manner set out in the Specification or, where not so set out, to the reasonable satisfaction of the Engineer and all work shall be carried out in accordance with such reasonable directions as the Engineer may give.

This is the same as clause 13.2 of MF/1 with 'Plant' substituted for 'Works'.

Vesting of plant

Clause 26.1 (ownership of plant) and clause 26.2 (marking of plant) are identical to clauses 37.1 and 37.2 of MF/1.

6.5 Express obligations and responsibilities of the contractor in MF/2

clause 2.5	– to request without delay that the engineer confirms instructions in writing
clause 3.2	– to be responsible for the acts, defaults and neglects of sub-contractors
clause 5.1	– to satisfy himself on all circumstances affecting the contract
clause 5.2	– to be responsible for interpretation of site data
clause 5.3	– to be responsible for incorrect information other than that provided in writing by the purchaser or the engineer
clause 7.1	– to enter into a formal contract agreement if so required by the purchaser
clause 8.1	– to provide a performance bond if so required by the purchaser
clause 9.1	– to treat details of the contract as confidential
clause 13.1	– to design, manufacture, deliver and test the plant, all within the time for completion
	– to made good defects during the defects liability period
clause 13.2	– to execute and manufacture the plant in accordance with the specification
	– to carry out work in accordance with the reasonable directions of the engineer
clause 13.3	– to be responsible for detailed design
	– to give notice of any disclaimer of responsibility for the purchaser's/engineer's design within a reasonable time of receipt
clause 14.1	– to submit a programme within 30 days of the letter of acceptance
clause 14.5	– to revise the programme if so required by the engineer
clause 14.6	– to expedite progress if so notified by the engineer
clause 15.1	– to submit drawings, etc for approval and to supply additional copies of approved drawings
clause 15.2	– to modify and resubmit for approval drawings, etc which the engineer disapproves
clause 15.5	– to provide drawings, etc showing how the plant is to be fixed and other relevant information

clause 15.6 – to supply operating and maintenance instructions prior to taking-over

clause 16.1 – to be responsible for errors and omissions in drawings, etc unless they are due to information supplied by the purchaser or the engineer

– to carry out remedial work necessitated by drawings, etc for which the contractor is responsible and to modify drawings, etc accordingly

clause 17.1 – to obtain permission for the engineer to inspect, examine and test on sub-contractor's/supplier's premises

clause 17.2 – to agree dates for testing with the engineer and to supply certified copies of test results if the engineer does not attend

clause 17.5 – to repair or replace rejected plant and to resubmit for testing

clause 18.1 – to apply for permission to deliver plant to site

clause 19.1 – to store, preserve, protect and insure the plant during periods of suspension

clause 19.5 – to make good deterioration, defects or loss occurring during suspension

clause 20.1 – to make good defects specified by the engineer before delivery

clause 21.2 – to carry out variations instructed by the engineer

clause 21.4 – to keep contemporary records of the cost of variations

clause 21.5 – to notify the engineer if any variation is likely to prevent or prejudice the contractor from fulfilling his obligations

clause 21.6 – to immediately proceed with variations on receipt of the engineer's instructions

clause 22.1 – to execute the contract so that the plant is delivered within the time for completion

clause 23.1 – to give notice of any claim for extension of time as soon as reasonably practicable after the delaying event

clause 23.3 – to consult with the engineer to mitigate delay and to comply with the engineer's instructions thereon

clause 24.1 – to pay damages for delay in completion

clause 25.2 – to be responsible for making good defects or damage appearing or occurring during the defects liability period

clause 25.8 – to make good defects appearing within three years of taking-over providing the defects were caused by 'gross misconduct'

clause 26.2 – to set aside and mark plant in which property has passed to the purchaser and to permit the engineer to inspect the same

clause 27.9 – to make application for the final payment certificate forthwith after completing defects liability obligations

clause 29.1 – to give notice of claims within 30 days of the circumstances arising and to submit full particulars of amounts claimed
clause 30.1 – to indemnify the purchaser against infringement of patents, etc
clause 31.1 – to be responsible for care of the plant until delivery
clause 31.2 – to make good loss or damage to the plant prior to delivery
clause 33.1 – to insure the plant against loss or damage until delivery
clause 33.2 – to apply insurance monies towards repair and replacement of the works
clause 33.3 – to effect insurances with an approved insurer and on approved terms
          – to procure the waiver of subrogation rights in respect of insurances in joint names.

6.6 Obligations of the vendor in MF/3

Form of tender

In the specimen form of tender the vendor offers to:

- supply the goods
- observe the provisions of the contract
- execute a formal agreement if so required
- furnish a performance bond if so required
- do extra work as ordered by the purchaser.

Compared with the form of tender in MF/2 insurance is missing but this follows the absence of any requirements on insurances in MF/3.

Form of agreement

The obligations of the vendor as stated in the specimen form of agreement are:

- to supply the goods
- to remedy defects.

Note that there is no express mention of design or delivery but both can reasonably be implied.

Obligations in the general conditions

MF/3 does not contain any statement of general obligations of the vendor.

Nor does it contain any statement to the effect that the various obligations of the vendor and the purchaser set out in the conditions are exclusive obligations.

Consequently the express list of obligations and responsibilities of the vendor which follows in section 6.7 is not necessarily, as with MF/1 and MF/2, an exclusive list of the contractor's obligations.

6.7 Express obligations and responsibilities of the vendor in MF/3

clause 3.2	– to supply to the purchaser within a reasonable time drawings and other particulars
clause 4.1	– to inspect and test the goods before delivery and to supply the purchaser with test result certificates
clause 4.2	– to carry out prescribed tests before delivery
	– to give the purchaser seven days notice of such tests
	– to supply certificates of test results if the purchaser fails to attend
clause 4.3	– to carry out repeat tests after failure
clause 5.2	– to pay the purchaser expenditure reasonably incurred in replacing reject goods
clause 6.1	– to deliver the goods to the place named in the contract or to load goods on to the purchaser's vehicle
clause 7.3	– to pay liquidated damages for delay in delivery
clause 7.4	– to pay the purchaser's reasonable expenditure in obtaining goods in place of undelivered goods
clause 8.1	– to store, protect, preserve and insure goods held back on the instructions of the purchaser
	– to use best endeavours to deliver such goods when so required
clause 9.1	– to repair or replace goods damaged in transit
clause 9.2	– to replace goods lost in transit
clause 11.1	– to make good or replace goods found to be defective within 12 months of delivery
clause 12.1	– to indemnify the purchaser against claims for patent infringements
clause 13.1	– to carry out variations as instructed by the purchaser
clause 13.2	– to advise the purchaser as soon as reasonably possible if any instruction or variation will affect the contract price
clause 13.3	– to notify the purchaser if any variation is likely to prevent or prejudice fulfilment of the vendor's obligations
clause 14.1	– to indemnify the purchaser against damage or injury to the purchaser's property or person.

6.8 *Obligations of the contractor in G/90*

G/90 does not contain any specimen forms of tender or agreement.

Not infrequently the forms of tender and agreement from the ICE Conditions of Contract are used with G/90 but they are not wholly suitable – not least because of the absence of any mention of design obligations.

Obligations in the general conditions

One of the difficulties of using G/90 is that obligations of the contractor and the purchaser are frequently put together in the same clause and there is no clear statement of the contractor's general obligations. And, as mentioned in Chapter 4, G/90 is silent on some important obligations such as design.

This, however is not fatal, since in G/90 unlike MF/1 and MF/2 there is nothing making the express obligations of the parties their exclusive obligations.

Section 6.9 which follows schedules the express obligations and responsibilities of the contractor in G/90. Comment is given below on those provisions not covered in other chapters. These are:

- clause 2   – contractor to inform himself
- clause 8(i)   – manner of execution
- clause 9(i)   – contractor's equipment, etc
- clause 9(ii)   – fencing and watching, etc
- clause 9(iii)   – electricity, water and gas
- clause 9(iv)   – lifting equipment
- clause 16(iii)   – facilities for the engineer
- clause 16(iv)   – facilities for other contractors
- clause 16(v)   – working hours
- clause 16(vi)   – safety
- clauses 17(i) to 17(vii) – vesting of plant and equipment
- clauses 20(i) and 20(ii) – contractor's responsibilities and workmen.

Clause 2 – contractor to inform himself fully

> The Contractor shall be deemed to have examined the Site, if access thereto has been available to him, and the General Conditions and Specification, with such schedules, drawings, and plans as are annexed thereto or referred to therein.

This clause does not go on to say that the contractor is deemed to have based his tender on his findings but that can be implied.

Clause 11 (underground works) gives entitlement to the contractor to recover extra cost in respect of general conditions which could not have reasonably been inferred from site inspection.

Clause 8(i) – manner of execution

The Works shall be performed in the manner set out and in all respects in compliance with the Specification or when not referred to in the Specification in accordance with the instructions in writing of the Engineer and in both cases shall be performed and completed to the reasonable satisfaction of the Engineer.

This is much the same as clause 13.2 of MF/1.

The clause appears to suggest that the engineer can only give instructions when the specification is silent but in effect it confirms that any departure from the specification is a variation.

Clause 9(i) – contractor's equipment, etc

Unless specific arrangements be made to the contrary the Contractor shall, at his own expense, provide all Contractor's Equipment and all materials, labour, haulage, and power necessary to execute and complete the Works.

This could be described as a general obligations clause and in MF/1 a similar sentence is found in clause 13.1 (general obligations).

The 'specific arrangements' referred to should be detailed in the special conditions.

Clause 9(ii) – fencing and watching, etc

The Contractor shall be responsible for the proper fencing, guarding, lighting, and watching of all the Works on the Site until taken over under Clause 28 (Taking Over) and for the proper provision during a like period of temporary roadways, footways, guards, and fences as far as the same may be rendered necessary by reason of the Works for the accommodation and protection of the owners and occupiers of adjacent property, the public, and others. No naked light shall be used by the Contractor on the Site otherwise than in the open air without special permission in writing from the Engineer.

This clause is virtually identical to clause 18.1 of MF/1.

Clause 9(iii) – electricity, water and gas

The Contractor shall be entitled to use for the purposes of the Works such supplies of electricity, water, and gas as may be available therefor on the Site and shall pay to the Purchaser for such use such sum as may be reasonable in the circumstances, and shall at his own expense provide any apparatus necessary for such use.

This is the equivalent of clause 18.2 (site services) of MF/1 in confirming that it is the contractor's obligation both to provide apparatus for using, and to pay for using, site services made available by the purchaser.

Clause 9(iv) – lifting equipment

The Purchaser shall at the request of the Contractor and for the execution of the Works operate any suitable lifting equipment belonging to the Purchaser that may be available on the Site and the Contractor shall pay a reasonable sum therefor. The Purchaser shall during such operation retain control of and be responsible for the safe working of the lifting equipment but shall not be responsible for any negligence of the Contractor.

The only obligation of the contractor here is to pay a reasonable sum for the use of the purchaser's lifting equipment.

Clause 16(iii) – facilities for the engineer

In the execution of the Works, the Contractor shall not authorize or purport to authorize any person other than his employees and Sub-contractors and their employees to come upon the Site, except by the written permission of the Engineer, but facilities to inspect the Works at all times shall be afforded to the Engineer and his representatives and authorized representatives of the Purchaser.

This clause has no direct equivalent in MF/1.

Its provisions are reasonable in principle but the requirement to afford facilities to inspect 'at all times' could perhaps be better expressed.

Clause 16(iv) – facilities for other contractors

The access to and possession of the Site referred to in Sub-Clause (i) hereof shall not be exclusive to the Contractor but only such as shall enable him to execute the Works. The Contractor shall afford to the Purchaser and to other contractors whose names shall have been previously communicated in writing to the Contractor by the Engineer every reasonable facility for the execution of work concurrently with his own.

The provisions in this clause are found in many standard forms – including MF/1 (in clause 18.4).

Usually, however the contractor is given an express right to payment in respect either of particular facilities or facilities beyond those deemed to be reasonable. In this case, however, there is no mention of payment.

Clause 16(v) – working hours

> The Purchaser shall give the Contractor facilities for carrying out the Works on the Site continuously during the normal working hours generally recognized in the district. The Engineer may, after consulting with the Contractor, direct that work shall be done at other times if it shall be practicable in the circumstances for work to be so done, and a sum for work so done shall be added to the Contract Price unless such work has, by the default of the Contractor, become necessary for the completion of the Works within the time fixed by the Contract, or, if no time be fixed within a reasonable time. Such sum shall be ascertained and determined in like manner to the valuation of variations under Clause 10 (Variations and Omissions).

This clause is similar to clause 19.1 of MF/1 but note that it does not commence 'Unless otherwise provided'. This suggests the contractor has a right to work during normal hours or to be paid the extra cost in other circumstances.

If normal working hours are not intended in the contract this clause should be amended.

Clause 16(vi) – safety

> The Contractor shall take full responsibility for the adequacy, stability and safety of his Site operations and installations and for the due satisfaction of his obligations under the Health & Safety at Work etc, Act 1974.

Somewhat unusually this clause refers to a specific statute.

It does not expressly mention safety requirements of the purchaser but if these apply they should be detailed in the special conditions.

Vesting of plant and equipment

The provision in clauses 17(i) to 17(vii) of G/90 on the vesting of plant and contractor's equipment are amongst the most comprehensive and extensive provisions on the subject found in any standard form.

The extent to which they are applicable however has to be seen in the light of comments earlier in this chapter on privity of contract and the difficulties of excluding legal rights of third parties.

In short the provisions in G/90 cover:

- clause 17(i)   – the date on which property in plant vests in the purchaser
- clause 17(ii)   – the vesting of the contractor's equipment on the site
- clause 17(iii) – reversion of vesting in the event of the purchaser's bankruptcy
- clause 17(iv)   – storage and responsibility of plant prior to delivery
- clause 17(v)   – the purchaser's rights in the event of termination of the contract for contractor's default or bankruptcy
- clause 17(vi)   – the contractor's obligation to protect the title of the purchaser
- clause 17(vii) – the provisions to be incorporated into sub-contracts in respect of vesting.

Contractor's representatives and workmen

Clause 20(i) and 20(ii) of G/90 are identical to clauses 17.1 and 17.2 of MF/1.

They require the contractor to employ a named representative on the site during working hours and to remove persons whom the engineer considers incompetent or negligent.

6.9 Express obligations and responsibilities of the contractor in G/90

clause 2	– to inform himself fully on the site, the general conditions and the specification
clause 3(i)	– to provide security for performance if required by the purchaser
clause 4(i)	– to submit drawings, etc for approval
clause 4(iv)	– to furnish information necessary to enable the purchaser to operate and maintain the works
clause 5(i)	– to be responsible for mistakes in drawings, etc unless due to incorrect information from the purchaser or the engineer
clause 6(i)	– to obtain consent to assignment
clause 6(ii)	– to obtain consent to sub-letting
clause 7(i)	– to indemnify the purchaser against claims for infringement of patents
clause 8(i)	– to perform the works to specification
	– to comply with engineer's instructions
	– to complete to reasonable satisfaction of the engineer
clause 8(ii)	– to submit a programme within the time stated in the appendix

clause 8(iii) – to vary the programme on instructions from the engineer

clause 8(iv) – to deliver plant on time by the expiration of the relevant delivery period

clause 8(v) – to complete the works by the expiration of the relevant time for completion

clause 9(i) – to provide the equipment for the works and all necessary materials, labour, etc

clause 9(ii) – to fence and guard the works on site until take-over

clause 9(iii) – to pay the purchaser for the use of services on site

clause 9(iv) – to pay for the use of the purchaser's lifting equipment

clause 10(i) – to carry out variations as directed by the engineer

– to notify the engineer if any variation is likely to involve a net addition or deduction of more than 15% from the contract price

clause 10(ii) – to notify the engineer of any variations likely to prevent or prejudice performance

clause 11 – to inform the engineer of proposed steps to deal with unforeseen ground conditions

clause 14(i) – to obtain permission for the engineer to inspect and test plant on supplier's/sub-contractor's premises

clause 14(ii) – to give the engineer notice when plant is ready for testing and to provide certified test results if the engineer fails to attend

clause 14(iii) – to provide assistance and resources for tests on the premises of the contractor or any sub-contractor

clause 15(i) – to obtain authorization for, and arrange delivery of plant to site and to be responsible for off-loading, etc

clause 15(vi) – to make good any deterioration or defect in the plant resulting from delayed delivery

clause 16(iii) – to afford facilities to the engineer and his representatives to inspect the works

clause 16(iv) – to afford reasonable facilities on the site to the purchaser and other contractors

clause 16(v) – to work outside normal working hours if so directed by the engineer

clause 16(vi) – to take full responsibility for the adequacy, stability and safety of all site operations

clause 17(i) – to ensure that title in goods passes to the contractor not later than delivery of the goods to the purchaser

clause 17(ii) – to be liable for loss or destruction of contractor's equipment unless through the fault of the purchaser

clause 17(iv) – to set aside and mark plant, property in which title has passed to the purchaser

clause 17(v) – in the event of termination to deliver such plant to the purchaser clause 17(vi)   – to take steps to ensure that the purchaser's title in plant is known to others dealing with the plant

clause 17(vii) – to incorporate equivalent provisions on the title of plant in sub-contracts

clause 18(ii)   – to be responsible for ensuring that dimensions and levels are correct and to set out the works

clause 18(iii) – to carry out the works to the reasonable satisfaction of the engineer

clause 19   – to proceed with the works in accordance with the engineer's instructions, etc

clause 20(i)   – to employ one or more named representatives on the site

clause 20(ii)   – to remove from the site persons who are in the opinion of the engineer incompetent or negligent

clause 21(i)   – to be responsible for care of the works until they are taken over

clause 21(ii)   – to make good loss or damage to the works caused by the excepted risks if so required by the purchaser

clause 21(iv)   – to indemnify the purchaser against loss or damage to third parties

clause 23(i)   – to insure the plant against loss, destruction or damage until delivery

                – to produce satisfactory evidence of insurance cover

clause 23(ii)   – to insure the works against loss or damage until taking-over

                – to apply any insurance monies to repair or replacement

clause 23(iii) – to insure against loss, damage or injury to third parties

                – to insure with an approved insurer on approved terms

clause 23(iv)   – to produce insurance policies and premium receipts as required by the purchaser

clause 24   – to remedy defects prior to taking-over as specified by the engineer

clause 25   – to give notice of any claim for extension of time without delay

clause 26   – to pay liquidated damages for late completion

clause 27(i)   – to give 21 days notice to the engineer of readiness for tests before completion

clause 27(v)   – to repeat tests which have failed

clause 27(vii) – to pay liquidated damages for low performance if specified and due

clause 28(iv)   – to carry out tests delayed by interference when so required

clause 28(v)   – to protect and preserve the works deemed to have been taken over

clause 29(i)   – to give notice of any claim made under clause 29(i) within a reasonable time after the event

clause 30(i)   – to be responsible for making good defects after taking over
clause 30(v)   – to carry out tests before completion on replacements and renewals if so required
clause 31(ii)   – to apply for interim payment certificates
clause 31(vi)   – to apply for final payment certificates
clause 31(xii) – to make good latent defects appearing within three years of taking over
clause 36   – to give notice to the engineer of problems with metrication.

Chapter 7

Assignment and sub-contracting

7.1 Introduction

In common with many other standard forms the model forms for plant contracts contain restrictions on the contractor's freedom to assign and sub-contract.

These restrictions are expressed in the model forms in clauses which are reasonably clear in their meaning as to their intended practical effect. Certainly ordinary users of the forms should find little difficulty in understanding their purpose.

But, as shown in the next section of this chapter, assignment and sub-contracting are complex legal issues which even eminent Lord Justices struggle to explain. A degree of caution is therefore recommended in dealing with these matters.

Responsibilities for sub-contractors

Sub-contractors fall into three main categories:

- domestic sub-contractors
- named or specified sub-contractors
- nominated sub-contractors.

Domestic sub-contractors are selected and appointed by the contractor to suit his own arrangements. Named, or specified, sub-contractors are selected by the engineer or purchaser but the terms of appointment are usually left to the contractor. Nominated sub-contractors are specified in the contract or their use is instructed by the engineer and the terms of their appointment, particularly with regard to price, are often outside the contractor's control.

Domestic sub-contractors are readily understood to be the responsibility of the contractor. And usually the same responsibility applies to named or specified sub-contractors although the case for this may not be as apparent. With nominated sub-contractors however there are differing views on whether responsibility for their performance should rest with the contractor or the purchaser. In the ICE forms of contract, the contractor takes responsibility; in the IChemE model forms the purchaser takes responsibility.

No express provision for nominated sub-contractors

The model forms for plant contracts avoid the problem to the extent that they make no express provision for nominated sub-contractors.

The general rule in MF/1, MF/2 and G/90 is that the contractor takes responsibility for sub-contractors. But this however is qualified by provisions that the contractor shall not have responsibility for work done by sub-contractors appointed under prime cost items unless the contractor approves the appointment. In effect, nominated sub-contracting is contemplated but it is not regulated.

This is arguably the worst possible state of affairs and engineers, in the interests of purchasers, should avoid using prime cost items and nominating under them if alternative ways of specifying the work can be arranged.

7.2 *Assignment and sub-contracting generally*

Definitions

Assignment can be defined as a transfer of title or interest in land, property or contractual rights. Sub-contracting can be defined as an arrangement to secure vicarious performance of contractual obligations.

The result may seem to be the same but in law the treatment of rights and obligations is different. And assignment and sub-contracting are quite different in their approach to the basic doctrine of privity of contract. Assignment is an exception to the doctrine of privity; sub-contracting gives no relief from the doctrine.

Assignment then is a process which confers legal rights and can include the transfer but not extinguishment of contractual obligations. Sub-contracting is no more than delegation of performance. It does not relieve the contractor of his obligations to the purchaser and it does not create any contractual rights or obligations between the purchaser and sub-contractor.

Assignment and sub-contracting distinguished

The courts have frequently had to consider whether the sub-contracting of obligations amounts to assignment. This is how the Master of the Rolls, Lord Greene, saw the matter in *Davies* v. *Collins* (1945), He said:

'In many contracts all that is stipulated for is that the work shall be done and the actual hand to do it need not be that of the contracting party himself; the other party will be bound to accept performance carried out by somebody else. The contracting party, of course, is the only party who remains liable. He cannot assign his liability to a sub-contractor, but his liability in those cases is to see that the work is done, and if it is not properly done he is liable. It is quite a mistake to regard that as an assignment of the contract; it is not.'

Benefits and burdens

Legal analysis of assignment and sub-contracting is frequently explained in terms of benefits and burdens. As a general rule of English law, benefits can be assigned but not burdens.

This is how Lord Justice Staughton in the case of *Linden Gardens Trust Ltd* v. *Lenesta Sludge Disposals Ltd* (1992) distinguished between the three basic concepts of novation, assignment and sub-contracting:

'(a) Novation
This is the process by which a contract between A and B is transformed into a contract between A and C. It can only be achieved by agreement between all three of them, A, B and C. Unless there is such an agreement, and therefore a novation, neither A nor B can rid himself of any obligation which he owes to the other under the contract. This is commonly expressed in the proposition that the burden of a contract cannot be assigned, unilaterally. If A is entitled to look to B for payment under the contract, he cannot be compelled to look to C instead, unless there is a novation. Otherwise B remains liable, even if he has assigned his rights under the contract to C.

Similarly, the nature and content of the contractual obligations cannot be altered unilaterally. If a tailor (A) has contracted to make a suit for B, he cannot by an assignment be placed under an obligation to make a suit for C, whose dimensions may be quite different. It may be that C by an assignment would become entitled to enforce the contract – although specific performance seems somewhat implausible – or to claim damages for its breach. But it would still be a contract to make a suit that fitted B, and B would still be liable to A for the price.

(b) Assignment
This consists in the transfer from B to C of the benefit of one or more obligations that A owes to B. These may be obligations to pay money, or to perform other contractual promises, or to pay damages for a breach of contract, subject of course to the common law prohibition on the assignment of a bare cause of action.

contractors'. However in *Jarvis Ltd* v. *Rockdale Housing Association Ltd* (1986) the Court of Appeal held that 'contractor' in the phrase 'unless caused by some negligence or default of the contractor' in the determination provisions of JCT 80 applied only to the contractor as main contractor and did not include nominated sub-contractors.

These two extracts from the judgment of Lord Justice Bingham explain the logic of the decision:

> 'I cannot accept the employer's submission that "the Contractor" in 28.1.3.4 is to be understood as including sub-contractors and their servants and agents. Both in this clause and elsewhere in the contract the draftsman has made express reference to sub-contractors, their servants and agents. The absence of any such reference here is not explained by the context, nor can it be dismissed as of no significance. Even an inexperienced draftsman could be relied upon to avoid such an error'.
>
> "The Contractor" can in my judgment be naturally and sensibly understood as referring to, in this case, John Jarvis Ltd, its servants and agents, through whom alone it can, as a corporation, act.'

It is clearly most important that contract draftsmen express their intentions with both clarity and consistency. The essential rule is – 'never change your wording unless you intend to change your meaning'.

For comment on the application of the *Jarvis* case to the model forms for plant contracts see section 7.4 below.

Removal of sub-contractors

Many standard forms, including MF/1, entitle the engineer to require the removal from site of any of the contractor's employees who are negligent or incompetent. Few forms, other than the ICE Design and Construct Conditions, expressly make similar provision for the removal of sub-contractors. Without such express power it is unlikely that the engineer's powers can properly be extended beyond applying to individuals to applying to firms.

Terms of sub-contracts

MF/1 and MF/2 do not attempt to regulate the terms on which sub-contracts should be let. Some contracts including the IChemE Green Book, GC/Works/1 (edition 3) and the Highways Agency Design and Build form do this – principally to ensure that the obligations of sub-contractors match those of the contractor.

MF/1 and MF/2 do contain model forms of sub-contracts but their use is not mandatory. G/90 has no model form of sub-contract but it does require in clause 17(vii) that reasonable steps be taken to incorporate provisions into sub-contracts relating to payment for plant before it is delivered.

7.3 Assignment in the model forms

MF/1 and MF/2

Clauses 3.1 of MF/1 and MF/2 are identical.

> The Contractor shall not assign the benefit of the Contract in whole or in part or any of his obligations under the Contract. A charge in favour of the Contractor's bankers of any monies due under the Contract, or the subrogation of insurers to the Contractor's rights shall not be considered an assignment.

By clause 49.1 of MF/1 (and 34.1 of MF/2) if the contractor does so assign the purchaser is entitled to terminate the contract.

The exception in clause 3.1 for a charge in favour of the contractor's bankers is narrow in commercial terms. Corresponding exemptions in other contracts usually have wider scope. On the face of it, assignments to factoring companies and the like are not permitted in MF/1 and MF/2.

The reference to subrogation in clause 3.1 is superfluous. Subrogation is a legal right of insurers and it is not ordinarily considered as assignment.

MF/3

MF/3 does not expressly cover assignment. Common law principles, therefore, apply.

G/90

Clause 6(i) applies:

> The Contractor shall not, without the consent in writing of the Purchaser, which shall not be unreasonably withheld, assign or transfer the Contract, the benefits or obligations thereof or any part thereof to any other person, provided that this shall not affect any right of the Contractor to assign, either absolutely or by way of charge, any moneys due or to become due to him, or which may become payable to him under the Contract.

G/90 does not expressly state that assignment without consent is a default entitling the purchaser to terminate. But such action could be regarded as a default within the scope of clause 12 in so far that it could be said to 'contravene the provisions of the Contract'.

The provision in clause 6(i) that the purchaser's consent shall not be unreasonably withheld might be seen to have the unintended effect that the contractor actually has an entitlement to assign – subject only to obtaining consent. For this reason the phrase is best omitted.

The exemptions in respect of monies due to the contractor are far wider than in MF/1.

7.4 Sub-contracting in MF/1 and MF/2

MF/1 and MF/2 have near identical provisions on sub-contracting.

Definitions

Clause 1.1.c (MF/1)

> 'Sub-Contractor' means any person (other than the Contractor) named in the Contract for any part of the Works or any person to whom any part of the Contract has been sub-let with the consent in writing of the Engineer, and the Sub-Contractor's legal successors in title, but not any assignee of the Sub-Contractor.

The reason for making 'Sub-Contractor' a defined term is not readily apparent and few other standard forms find it necessary.

However the clause does confirm that sub-contractors whether named in the contract or selected by the contractor are intended to have the same contractual status.

That apart, the only significance of the defined term is in those clauses of the contract where it is used. In MF/1 these are:

- clause 17.3 – returns of labour
- clause 21.1 – extraordinary traffic
- clause 23.3 – services for tests and inspection
- clause 26.1 – defects before taking-over
- clause 33.2 – delays by sub-contractors
- clause 36.9 – limitation of liability for defects
- clause 43.5 – injury to persons and damage
- clause 43.6 – accidents or injury to workmen
- clause 47.5 – insurance against accidents to workmen
- clause 53.1 – sub-contractors, servants and agents.

The danger of these various references in the text of MF/1 (and MF/2 has similar references) is it casts doubt on what is meant by the term 'Contractor' when used alone. Does 'Contractor' then include for sub-contractors? See the case of *Jarvis* v. *Rockdale* mentioned earlier in this chapter.

A further complication is that the definition of sub-contractor includes only those named or those consented to in writing by the engineer. But it is clear from clause 3.2 that the contractor is not required to obtain consent to all sub-contractors.

Precisely how these 'non-defined' sub-contractors fit into the clauses listed above is not clear. For example, where clause 33.2 permits extensions of time for certain delays affecting 'Sub-Contractors' as defined does it by implication exclude similar delays affecting 'non-defined' sub-contractors?

Consents and responsibility

Clause 3.2 (MF/1)

> Except where otherwise provided by the Contract the Contractor shall not sub-contract any part of the Works without the prior consent of the Engineer.
>
> The Contractor shall however not require such consent to place contracts for minor details nor for purchases of materials nor for any part of the Works of which the manufacturer or supplier is named in the Contract.
>
> The Contractor shall be responsible for the acts, defaults and neglects of any Sub-Contractor, his agents, servants or workmen as fully as if they were the acts, defaults or neglects of the Contractor, his agents, servants or workmen.

'Except where otherwise provided'

These words presumably relate to the provisions which follow in the clause; to any sub-contractors instructed by the engineer under a prime cost item; and to sub-contractors named by the contractor in his tender.

'Without the prior consent'

Note that under clause 1.4 any consent required shall not be unreasonably withheld.

In clause 49.1 sub-letting the whole of the works without the consent of the purchaser is listed as a default entitling the purchaser to terminate the contract. Whether sub-contracting any part of the works without the prior consent of the engineer is a default of similar consequence is open to argument. Most probably it is not – although that may not be what is intended.

'Responsible for the acts, defaults and neglects of any Sub-Contractor'

These words appear to take the responsibility of the contractor for his sub-contractors beyond mere performance under the contract. They raise difficult questions as to the responsibility of sub-contractors for the negligence of their own employees.

Contractors might well be excused for finding the final provision of clause 3.2 unacceptable but the provision may not be as far reaching as the words suggest. Responsibility in the context of clause 3.2 is responsibility only of the contractor under the contract to the purchaser, not to others who might be affected by the negligence of sub-contractors.

Contractor not responsible

Clause 5.6 (MF/1)

> The Contractor shall have no responsibility for work done or Plant supplied by any other person in pursuance of directions given by the Engineer under this Clause unless the Contractor shall have approved the person by whom such work is to be done or such Plant is to be supplied and the Plant, if any, to be supplied.

The implications for the purchaser if the contractor is not responsible for all the work done under the contract are potentially immense.

In order to protect his position in such circumstances the purchaser would be well advised to seek guarantees or collateral warranties from those persons or firms outside the scope of the contractor's responsibility. For further comment on this clause and clause 5.6 of MF/2 see Chapter 18.

Sub-contractor's servants and agents

Clause 53.1 (MF/1)

> It is expressly agreed that no servant or agent of the Contractor nor any Sub-Contractor shall in any circumstances whatsoever (with the exception of liability for death or personal injury caused by wilful negligent acts or omissions) be under any obligation, responsibility or liability to the Purchaser for or in respect of any loss, damage or injury of whatsoever kind and howsoever arising. Without prejudice to the generality of the foregoing every limitation and exclusion of liability of the Contractor contained in the Conditions shall also extend to protect every such servant, agent or Sub-Contractor. For the purposes of this Clause the Contractor is or shall be deemed to be acting as agent or trustee on behalf of and for the benefit of all persons who are or who may from time to time become servants, agents or Sub-Contractors as aforesaid

and to such extent all such persons shall be or be deemed to be parties to the Contract.

The intention of this clause is to protect sub-contractors, servants and agents of the contractor from legal action by the purchaser. That is it attempts to fix the purchaser's remedies against the contractor and no one else.

Whether or not in law it can be successful is another matter. It is questionable whether a sub-contractor sued for negligence by the purchaser could rely on this clause as a defence – not least because the sub-contractor is not a party to the contract in which the clause appears.

For further comment see Chapter 14, section 3, and the case of *Southern Water Authority* v. *Lewis & Duvivier* (1984).

7.5 *Sub-contracting in MF/3*

MF/3 has no provisions on sub-contracting and it is open to the vendor to sub-contract as he thinks fit subject only to the common law rules on personal selection.

7.6 *Sub-contracting in G/90*

Definition

Clause 1 – definition of terms (part)

> 'Sub-Contractor' shall mean any person (other than the Contractor) named in the Contract for any part of the Works or any person to whom any part of the Contract has been sub-let with the consent in writing of the Engineer, and the legal representatives, successors, and assigns of such person.

This is much the same definition as in MF/1 but in this case 'assigns' remain sub-contractors whereas in MF/1 'assignees' are excluded.

It is not clear why 'assigns' should be regarded as sub-contractors because as far as the purchaser is concerned they are likely to be only the recipients of benefits and not the carriers of burdens.

The clauses of MF/2 which expressly refer to 'Sub-Contractors' as defined are:

- clause 14(iii) – inspection, testing and rejection
- clause 17(vi) – vesting of plant
- clause 24   – defects prior to taking over
- clause 25   – extension of time for completion.

Consents

Clause 6(ii) – sub-letting

> The Contractor shall not, without the consent in writing of the Engineer, which shall not be unreasonably withheld, sub-let the Contract or any part thereof, or make any sub-contract with any person or persons for the execution of any portion of the Works but the restriction contained in this clause shall not apply to sub-contracts for materials, for minor details, or for any part of the Works of which the makers are named in the Contract. Any such consent shall not relieve the Contractor from his obligations under the Contract.

Sub-letting without consent is not a specified default in clause 12 of G/90 but as with assignment without consent it would be caught by the default of contravening the provisions of the contract.

'Named in the Contract'

These words would apply to firms named either in the specification or in the contractor's tender proposals.

Contractor not responsible

Clause 32(iii)

> The Contractor shall have no responsibility for work done or Plant supplied by any other person in pursuance of directions given by the Engineer under this clause unless the Contractor shall have approved the person by whom such work is to be done or such Plant is to be supplied and the Plant, if any, to be supplied.

See the comments in section 7.4 of this chapter on the similar provision in clause 5.6 of MF/1.

7.7 Model forms of sub-contract

MF/1 and MF/2 both contain model forms of sub-contract which aim to place, as far as practicable, the provisions of the sub-contract 'back to back' with the provisions in the main contract.

They do this with a good measure of success and accordingly the comments in this book on the provisions in MF/1 and MF/2 generally apply to the provisions in the forms of sub-contract.

A few points are worth adding here in relation to payments, insurances and arbitration.

Payments

There are no payment schemes whatsoever in the model forms of sub-contract. The scheme for each sub-contract must be set out in a payment schedule.

This is a better arrangement than trying to match the fall-back provisions of MF/1 and MF/2 because it leaves the parties to the sub-contract with greater flexibility in their commercial arrangements.

Insurances

The model forms of sub-contract require the contractor to include sub-contractors as co-insured under policies for insurance of the works. The effect of this is that claims fall on the contractor's insurance even in the event of the sub-contractor's fault.

In some circumstances, depending upon the amount of work being undertaken on site by sub-contractors, it may be in the interests of the contractor to require the sub-contractor to carry his own insurance for the works as well as other insurances which will be specified in the appropriate schedule.

Arbitration

The arbitration clauses of the model forms of sub-contract provide that a sub-contract dispute which is substantially the same as a main contract dispute should be joined with the main contract dispute in arbitration.

In principle this is an excellent idea but it overlooks the point that the consent of the purchaser is required to such an arrangement. And since there is nothing in the main contracts requiring the purchaser to give his consent it is left for the purchaser in each case to decide whether or not it suits his case to allow the sub-contractor to be joined in the arbitration.

Chapter 8

Administration and supervision

8.1 Introduction

It is rarely enough, in contracts for works of technical complexity, that the contract should do no more than set out the rights and obligations of the parties. There are other questions to be addressed:

- who will administer the contract?
- who will supervise the works?
- who will value the works?
- who will issue certificates?

In short, how will the contract be managed and who will represent the parties?

The engineer

In plant, process and civil engineering contracts the title 'engineer' has been used for many years to describe the person or firm appointed to manage the contract. Manage, that is, in the administrative sense described above. In the operational sense the title 'contracts manager' usually is applied to a key member of the contractor's organization.

However there is some evidence that the title 'engineer' is losing favour and in recent contracts other titles such as 'project manager' and 'employer's representative' are used. These can be seen as emphasizing either managerial involvement in the running of the contract or the function of the contract administrator in looking after the employer's (or purchaser's) interests. Whether such changes do anything to improve the operation of the contract is questionable.

In MF/1, MF/2 and G/90 the title 'engineer' remains in use; and the 'engineer' in these contracts it must be admitted has something of a dual role to play. The engineer does represent, or act as the purchaser's agent, in exercising some functions – such as issuing variations or ordering suspensions. But in carrying out other functions which involve the exercise of discretion – such as valuing and certifying – the engineer is required to act fairly and hold the balance between the parties.

Independence of the engineer

Whether or not it can ever be truly said that the engineer is independent in so far that his salary or his fees are paid by the purchaser it is essential that the engineer is given the freedom to exercise his professional judgment. And one aspect of this which is recognized in some plant and process contracts is that the purchaser should not be able to change the appointed engineer without the consent of the contractor.

Another aspect, which finds a place in MF/1 and MF/2, is that if the purchaser imposes restrictions on the authority of the engineer, such restrictions should be made known to the contractor at the time of tender.

Responsibility for the engineer

The point is made very clearly in the IChemE model forms that the purchaser is responsible for the acts, omissions and defaults of the engineer. Consequently any failure of performance by the engineer amounts to a breach of contract by the purchaser.

The position is arguably the same under plant and construction contracts but it is not expressed with equal clarity. This might oblige the contractor to rely on an implied term that the purchaser is responsible for the performance of the engineer.

But, as stated many times in this book, there is a potential problem with implied terms in MF/1 and MF/2 because of the exclusive remedy provisions. And this does cast some doubt on how far, under these model forms, the purchaser's responsibility for the performance of the engineer extends.

Engineer to be a person or a firm

There has been much debate in professional circles on whether the engineer to a contract should be a named individual or whether it is acceptable that the engineer should be a firm. The trend in recent contracts is to allow a firm to carry the title and broad responsibility (perhaps with professional indemnity insurances in mind) but to require an individual to be named to act as engineer. Some contracts require the named individual to be a chartered engineer.

The model forms for plant contracts do permit the engineer to be a firm but have no requirements for naming individuals.

Finality of the engineer's decisions

It is crucial to the operation of a contract to know whether the engineer's decisions and his certificates have any finality or whether they can be challenged by one or both parties and reversed in arbitration.

MF/1, MF/2 and G/90 all aim toward finality – subject to challenge within specified timescales. But there is a problem in G/90 in that the valuation powers of the engineer (if they exist) are implied rather than expressed and finality in such circumstances is a dubious concept.

8.2 Role of the engineer generally

The engineer to a contract has four possible functions to perform:

- designer
- supervisor/inspector
- contract administrator/certifier
- representative of the purchaser.

Some of these functions may be expressed in the contract as duties and some as powers.

Duties and powers

Duties are usually indicated by the phrase 'the Engineer shall'; powers by the phrase 'the Engineer may'. The phrase 'the Engineer shall be entitled to' is most likely to indicate a power.

Failure by the engineer to perform a duty is, in most contracts, a breach of contract for which the purchaser is liable in damages to the contractor. But that liability is, of course, dependent upon the contractor being able to establish proof of loss. And not all breaches of contract lead to loss so it is not enough for the contractor simply to prove that the engineer has failed to perform his duties.

Time for the exercise of duties and powers

Unless the contract specifies the time allowed to the engineer for the exercise of various duties and powers the engineer should always endeavour to act within a reasonable time. Failure to do so could again put the purchaser in breach of contract.

As to what is a reasonable time – that is a matter of fact to be determined on the circumstances of each case. But factors which might be taken into account could include:

- the complexity of the matter
- the urgency of the matter
- the notice given
- the contractor's progress
- the contractor's programme
- the engineer's resources.

For further comment on this see 'The time for ordering variations' in section 16.5 of Chapter 16.

Exercise of discretion

Some contracts expressly require the engineer to act impartially in exercising his discretion. Others such as MF/1 and MF/2 require the engineer to act fairly. There is a slight distinction between acting impartially and acting fairly but it is not particularly important.

More important is the point that whether or not the contract expressly requires the engineer to act fairly or impartially there is a common law duty to do so. This is how Lord Reid made the point in the case of *Sutcliffe* v. *Thackrah* (1974). He said:

'It has often been said, I think rightly, that the architect has two different types of function to perform. In many matters he is bound to act on his client's instructions, whether he agrees with them or not; but in many other matters requiring professional skill he must form and act on his own opinion ...

... The building owner and the contractor make their contract on the understanding that in all such matters the architect will act in a fair and unbiased manner and it must therefore be implicit in the owner's contract with the architect that he shall not only exercise due care and skill but also reach such decisions fairly holding the balance between his client and the contractor.'

Personal liability of the engineer

The engineer will, under the usual rules of law, be liable to the purchaser under his contract of engagement for breaches of duty. For example, it was held in the case of *Sutcliffe* v. *Thackrah* mentioned above that an employer could recover his loss from his architect who had negligently over-certified prior to the contractor going into liquidation. The architect's defence that he was acting in an arbitral capacity when certifying was not accepted by the court.

Many legal commentators felt that a similar remedy would be open to a contractor as an action in negligence when the engineer had caused loss by under-certifying. But the Court of Appeal in *Pacific Associates* v. *Baxter* (1988) rejected such a claim holding that the engineer had no duty of care in respect of economic loss to the contractor and that the arbitration clause in the contract provided the contractor's remedy. In so far as this decision is applicable the engineer is effectively protected from direct claims by the contractor arising out of alleged failure to carry out duties or to exercise discretion fairly and impartially.

However, engineers may well find some cause for concern, and contractors some cause for celebration in a recent decision of the Supreme Court of Canada in the case of *Edgeworth Construction Ltd* v. *N. D. Lea & Associates Ltd* (1993). The court held that engineers preparing tender documents owed a duty of care to tenderers and were liable for losses suffered by the contractor as a result of negligent misstatements.

8.3 Delegation and assistants

The engineer to the contract is not expected to perform all of his functions personally and most contracts make provision for the appointment of assistants and the delegation of authority.

Nature of delegation

It is a matter for debate whether powers which are delegated reside solely with the person to whom they are delegated unless and until revoked, or whether they are shared with the original holder of the powers.

There are two views on this but practical considerations suggest that a power should not be exercised simultaneously by different persons. However if the correct view is that delegated powers are shared powers this would not diminish the effectiveness of actions taken by one of the persons using his delegated power.

Authority to delegate

Authority to delegate will often be stated in the contract – perhaps with requirements on written notice and limitations in respect of certain matters. Some contracts permit general delegation; others require delegated powers to be individually stated.

It is frequently quite difficult to tell from the wording of contracts –

because of such phrases in the definitions clauses as 'the singular includes the plural' – whether the engineer can delegate to one or more persons. Some contracts even seem to contemplate the delegation of the same power to more than one person – but the practical difficulties which this can cause are only too obvious.

Assistants

Assistants and clerks of works play a vital role in the management and supervision of contracts. The key question is – what powers and duties do they possess?

The answer to this has to be found in the contract. Not uncommonly, unless they are given delegated powers, assistants and clerks of works are restricted to watching and supervising.

Powers to reject and condemn which are frequently assumed by assistants and clerks of works more often than not lack proper authority. But contractors, particularly those who use the engineer's staff to supervise and provide quality control for their operations do not always object to this.

8.4 *Role of the engineer in MF/1*

Clause 1.1.d – Definition of the engineer

> 'Engineer' means the person appointed by the Purchaser to act as Engineer for the purposes of the Contract and designated as such in the Special Conditions or, in default of any appointment, the Purchaser.

The 'person' suggests an individual. But see clause 1.2 which states that words importing persons shall include firms and other organizations having legal capacity.

The intention of MF/1 is that the contractor should know at the time of tender the name of the engineer – this is considered to be a matter of commercial importance. Hence the requirement that the engineer should be designated in the special conditions.

Failure to so designate raises the question – can the contractor refuse to accept an engineer notified to him in some other way – most obviously by letter after acceptance of tender? The answer is probably – yes. Clause 2.8 clearly restricts the freedom of the purchaser to select the engineer as he chooses once the contract is made.

And the phrase 'in default of any appointment, the Purchaser' can also be taken as a restriction on the ability of the purchaser to make a late appointment. But perhaps it is necessary to distinguish between 'appointing' an engineer and 'designating' him in the special conditions. The clause does not actually say 'in default of such designation the purchaser shall be the engineer'.

What the clause does in effect say, and this is quite useful, is that the contract can operate without an engineer because in such circumstances the purchaser is empowered to act as the engineer. This is not without its difficulties but it is not a wholly unusual arrangement (see MF/3) and it does preserve the essential administrative arrangements of the contract in the event that there is no engineer.

Clause 1.1.e – definition of engineer's representative

'Engineer's Representative' means any assistant of the Engineer appointed from time to time to perform the duties delegated to him under Clause 2 (Engineer and Engineer's Representative) hereof.

A common question here is – can there be more than one engineer's representative? Or more to the point – can more than one person perform delegated duties? The words 'any assistant' suggest the singular but see clause 1.3 which states that words importing the singular also include the plural.

The answer is, perhaps, to be deduced from clause 2 which in its references to the engineer's representative makes no mention of assistants or other persons entitled to perform delegated duties. But for practical considerations there will frequently have to be more than one person assisting the engineer and it is therefore suggested that there can be more than one appointed 'representative'.

Neither clause 1.1.e nor clause 2 states how the contractor is to be informed of the appointment of representatives nor whether it is the engineer or the purchaser who is to make such appointments.

Good practice requires the engineer or the purchaser to write to the contractor at commencement with a list of names of assistants and a schedule of their duties.

Clause 2.1 – engineer's duties

The Engineer shall carry out such duties in issuing certificates, decisions, instructions and orders as are specified in the Contract.

If the Engineer is required, under the terms of his appointment by the Purchaser, to obtain the prior specific approval of the Purchaser before exercising any of his duties under the Contract, particulars of such requirements shall be set out in the Special Conditions.

The requirement that the engineer 'shall' carry out his duties makes it a potential breach of contract for which the purchaser is liable if he does not. However, in MF/1, the contractor's remedies are subject to clause 44.4 (exclusive remedies) and may not be as extensive as in other contracts with similar wording.

Prior approval provisions of the type in clause 2.1 are most commonly taken in other contracts to apply to limitations on the value of variations which the engineer can order without the purchaser's consent. Such limitations, if they exist, should be set out in the special conditions of MF/1 notwithstanding the overall 15% limit in clause 27.

Where MF/1 differs from other contracts is in the words 'exercising any of his duties'. Normally the words used are 'exercising his authority'. There is a significant difference between the two because whilst the prior approval requirements can quite rightly be applied to the engineer's powers it is questionable as to whether they should be applied to his duties. The engineer owes it to both parties to carry out his duties not just to the purchaser.

Clearly if the purchaser does require the engineer to obtain prior approval before carrying out any of his duties the contractor should be informed in the special conditions.

Clause 2.1 does not directly address the question of whether the contractor is obliged to ascertain whether the engineer has obtained any specified prior approval before the contractor acts on the engineer's instructions. Some contracts do address this and they usually say that the contractor is entitled to assume that the engineer has obtained his necessary approvals before giving instructions. In MF/1 this perhaps can be implied from clause 2.4 which requires the contractor to proceed in accordance with the engineer's instructions.

Schedules of the express duties and powers of the engineer under MF/1 are given at the end of this section.

Clause 2.2 – engineer's representative

> The Engineer's Representative shall be responsible to the Engineer and shall watch and supervise the Works, and test and examine any Plant or workmanship employed in connection therewith.
>
> The Engineer's Representative shall have only such further authority as may be delegated to him by the Engineer under Sub-Clause 2.3 (Engineer's Power to Delegate).

Without delegated powers the engineer's representative is not much more than the eyes and ears of the engineer.

MF/1 does refer to the engineer's representative in a few clauses of practical effect:

- clause 18.1 – fencing, watching and lighting
- clause 18.4 – opportunities for other contractors
- clause 19.2 – hours of work.

But these only serve to illustrate that elsewhere where the contract refers to the engineer then the engineer's representative can only act with delegated powers.

The wording of clause 2.2 clearly implies that the engineer's representative is site based. This can present difficulties in large organizations where office-based project engineers are required to act in the administrative capacity of the engineer and perhaps to value and issue certificates. On strict interpretation it is probably inappropriate for such persons to be named as the engineer's representative although the practice is quite common.

Clause 2.3 – engineer's power to delegate

The Engineer may from time to time delegate to the Engineer's Representative any of his duties and he may at any time revoke such delegation.

Any delegation or revocation shall be in writing. The Engineer shall furnish to the Contractor and to the Purchaser a copy of any such delegation or revocation. No such delegation or revocation shall have effect until a copy thereof has been delivered to the Contractor.

Any written decision, instruction, order or approval given by the Engineer's Representative to the Contractor in accordance with such delegation shall have the same effect as though it had been given by the Engineer. If the Contractor disputes or questions any decision, instruction or order of the Engineer's Representative he may refer the matter to the Engineer who shall confirm, reverse or vary the decision in accordance with Sub-Clause 2.6 (Disputing Engineer's Decisions, Instructions or Orders).

The engineer is given wide freedom to delegate under MF/1 but it is questionable whether it is good practice to delegate important decision making and certification duties. Some contracts expressly disallow this.

In any event the phrase 'any of his duties' cannot be taken literally since this would entitle the engineer to delegate his reviewing role under clause 2.6 which cannot possibly be intended.

As to whether the engineer can delegate to a person other than the engineer's representative see the comment on clause 2.8 below.

Clause 2.4 – engineer's decisions, instructions and orders

The Contractor shall proceed with the Works in accordance with the decisions, instructions and orders given by the Engineer in accordance with the Contract.

This is a straightforward but very important provision. It requires the contractor to comply with instructions, etc given by the engineer – an obligation which is surprisingly not found in every contract.

Clause 2.5 – confirmation in writing

The Contractor may require the Engineer to confirm in writing any decision, instruction or order of the Engineer which is not in writing. The Contractor shall make such request without undue delay. Such a decision, instruction or order shall not be effective until written confirmation thereof has been received by the Contractor.

This clause does not positively require all instructions, etc to be in writing. The words 'The Contractor may' give the contractor an option – to act on an oral instruction or to decline to act until the instruction is in writing.

However the contractor would be well advised to require all instructions to be put in writing, not only for the record, but to avoid any argument that the last sentence of the clause invalidates all oral instructions.

Clause 2.6 – disputing engineer's decisions, instructions and orders

If the Contractor by notice to the Engineer within 21 days after receiving the decision, instruction or order of the Engineer in writing or written confirmation thereof under Sub-Clause 2.5 (Confirmation in Writing), disputes or questions the same, giving his reasons for so doing, the Engineer shall within a further period of 21 days by notice to the Contractor and the Purchaser with reasons, confirm, reverse or vary such decision, instruction or order.

If either the Contractor or the Purchaser disagrees with such decision, instruction or order as confirmed, reversed or varied he shall be at liberty to refer the matter to arbitration within a further period of 21 days. In the absence of such a reference to arbitration within the said period of 21 days such decision, instruction or order of the Engineer shall be final and binding on the parties.

This clause is apparently designed to set a procedure which acts as a condition precedent to arbitration. So much can be gathered from clause 52.1 (notice of arbitration) which states that a dispute on the engineer's decisions, etc shall not be referred to arbitration except in accordance with clause 2.6.

It would be better therefore if clause 2.6 was clearer in stating that the contractor 'shall' refer any disputed decision, etc to the engineer within 21 days if he wishes to avoid that decision becoming final and binding.

Note that only the contractor can commence the dispute procedure under clause 2.6 by challenging the decisions, etc of the engineer. This is in contrast to many other standard forms which allow the purchaser the same right. However, clause 2.6 may not apply to disputes on the engineer's certificates but see section 20.4 of Chapter 20 for further comment on this important matter.

In simple form the clause 2.6 procedure is as follows:

● engineer's decision, instruction or order
   ↓
● contractor to dispute within 21 days (see the comment below)
   ↓
● engineer to confirm, reverse or vary within 21 days
   ↓
● contractor or purchaser to refer to arbitration within 21 days
   <u>OR</u>
● the decision, instruction or order becomes final and binding.

There seems to be little doubt from the wording of clause 52.1 that if the contractor fails to dispute a decision, instruction or order within 21 days such a decision, etc becomes final and binding.

Clause 2.7 – engineer to act fairly

Wherever by the Conditions the Engineer is required to exercise his discretion:
– by giving his decision, opinion or consent
– by expressing his satisfaction or approval
– by determining value
– or otherwise by taking action which may affect the rights and obligations of either of the parties
he shall exercise such discretion fairly within the terms of the Contract and having regard to all the circumstances.

General comment has been made on the need to exercise discretion fairly earlier in this chapter.

All that needs to be said here is that the words 'fairly within the terms of the Contract' should emphasize to the engineer that 'fairness' is not to be construed as rewriting the contract so as to give it better balance between the parties. However harsh the terms of the contract, they must still be applied for the engineer to act 'fairly'.

The phrase 'having regard to all the circumstances' may seem to introduce wider boundaries to what is fair but it is suggested that the phrase means no more than that the engineer should ensure that he is acquainted with all the facts.

Clause 2.8 – replacement of the engineer

The Purchaser shall not appoint any person to act with the Engineer or in replacement of the Engineer without the Contractor's prior consent.

Comment on the replacement aspect of this clause has been made under clause 1.1.d (definition of the engineer) earlier in this chapter.

As to the meaning of the phrase 'to act with the Engineer' this is something of a mystery. It seems unlikely that it is intended to apply to the appointment of the engineer's representative – since such a meaning would give the contractor complete control over the appointment. It may however be intended to apply to delegation of the engineer's powers to persons other than the engineer's representative. It would not be unreasonable for the contractor to have some say in this. It may also be intended to apply to other persons appointed by the purchaser such as an expeditor whose role it is to monitor or in any way control the progress of the contract.

Another problem area of the clause is what happens if the engineer designated in the special conditions dies, retires or otherwise leaves the service of the purchaser. On the face of it, the purchaser has no option but to consult with the contractor on his replacement.

Express duties of the engineer under MF/1

clause 2.1	– to carry out such duties as are specified in the contract
clause 2.3	– to copy details to the contractor and the purchaser of any delegation or revocation of powers
clause 2.3	– to rule on disputes involving exercise of the engineer's representative's powers
clause 2.5	– to confirm in writing any decisions, instructions or orders as required by the contractor
clause 2.6	– to respond within 21 days to any dispute or question on his decisions, instructions or orders
clause 2.7	– to exercise discretion fairly
clause 14.6	– to notify the contractor if the rate of progress is too slow to meet the time for completion
clause 15.1	– to approve or disapprove drawings and the like submitted for approval and to identify by signature or otherwise items which are approved
clause 21.3	– to certify deductions from the contract price in respect of extraordinary traffic claims resulting from the negligence of the contractor
clause 23.2	– to give the contractor 24 hours notice of his intention to attend tests or inspections

clause 23.4	– to issue certificates when plant has passed test or inspections
clause 27.5	– to give reasonable notice to the contractor when ordering any variations
clause 29.2	– to issue a taking-over certificate and to certify the date on which the works passed tests for completion
clause 29.4	– to certify deductions from the contract price in respect of outstanding work done by the purchaser
clause 31.1	– to issue a taking-over certificate on the application of the contractor when the contractor is prevented from tests on completion
clause 31.2	– to certify the additional costs of carrying out tests on completion during the defects liability period following interference with such tests
clause 33.1	– to grant such extension of time as is reasonable following notice of claim by the contractor
clause 35.7	– to compile and evaluate with the contractor the results of performance tests (this duty can be undertaken by the purchaser)
clause 36.3	– to notify the contractor of defects appearing or damage occurring during the defects liability period (this duty can be undertaken by the purchaser)
clause 38.2	– not to withhold consent to removal of contractor's equipment not required for the execution of the works
clause 39.3	– to issue an interim certificate of payment within 14 days of application
clause 39.11	– to issue a final certificate of payment within 30 days after receiving application
clause 46.4	– to certify the contract value prior to the date of termination in the event of *'force majeure'*
clause 49.2	– to certify the termination value in the event that the purchaser terminates the contract because of the contractor's default
clause 51.3	– to certify the termination value and the contractor's expenditure and loss in the event of termination following the purchaser's default.

Express powers of the engineer under MF/1

clause 2.3	– to delegate powers to the engineer's representative and to revoke such delegation
clause 5.4	– to direct on the use of provisional sums
clause 5.5	– to direct on the use of prime cost items

clause 13.2   – to give reasonable directions on the manner of execution of work

clause 14.5   – to order revisions of the programme if progress does not match the programme

clause 15.1(b) – to require drawings and details to be submitted for approval

clause 15.4   – to inspect all drawings of any part of the works

clause 17.2   – to object to the presence on the works of any representative or person employed by the contractor who is guilty of misconduct or is incompetent or negligent

clause 17.3   – to require periodic returns of staff and labour

clause 18.4   – to give instructions on opportunities for other contractors

clause 21.2   – to give instructions in respect of the transport of special loads

clause 23.1   – to inspect, examine and test on the contractor's premises

clause 23.5   – to reject plant which is not in accordance with the contract

clause 25.1   – to suspend the progress of the works at any time and to give instructions on protecting and securing the works

clause 25.5   – to lift the suspension and instruct the contractor to proceed

clause 27.2   – to issue variations to the works

clause 28.3   – to give instructions on delayed tests

clause 28.5   – to reject the works or any section which fails to pass the tests on completion

clause 31.2   – to order tests during the defects liability period following interference

clause 33.3   – to give instructions to overcome or mitigate the consequences of delay

clause 35.2   – to carry out performance tests on behalf of the purchaser

clause 35.3   – to order the cessation of performance tests to avoid damage

clause 35.5   – to notify the contractor if the purchaser requires postponement of adjustments or modifications to the works

clause 36.8   – to order the contractor to search for defects

clause 37.2   – to give instructions on setting aside and the marking of plant and to withhold any interim certificate until plant is so marked and set aside

clause 39.6   – to correct or modify certificates

clause 41.1   – to request further particulars of claims.

8.5 The engineer in MF/2 and the purchaser in MF/3

MF/2

Clauses 1.1.d and 2.1–2.8 of MF/2 are identical to the similarly numbered provisions of MF/1.

There are slight differences between MF/2 and MF/1 in the duties and powers of the engineer as expressed in other clauses but these merely reflect the basic differences in the work content between the two model forms.

MF/3

There is no provision for the appointment of an engineer or any other 'independent' contract administrator in MF/3. The form relies on the purchaser to administer the contract.

In practice an engineer does often take on the burden of contract administration but he does so as the agent of the purchaser and his legal liabilities need to be considered accordingly.

8.6 Role of the engineer in G/90

Definition of the engineer

Clause 1 of G/90 contains the following definition:

> The 'Engineer' shall mean ... or the person for the time being or from time to time notified in writing by the Purchaser to the Contractor as the Engineer for the Contract, or in default of any notification the Purchaser.

This is a more flexible definition from the purchaser's viewpoint than that in MF/1. The purchaser has the freedom to replace the engineer as he thinks fit and he does not require the contractor's consent to do so.

Engineer's supervision

Clause 18(i):

> After the tender has been accepted by the Purchaser, all instructions and order to the Contractor shall, except as herein otherwise provided, be given by the Engineer.

The standard printed form has a misprint in the final word of this clause – stating 'Purchaser' instead of 'Engineer'. This was corrected by an erratum note issued by WSA in January 1991.

The clause confirms the authority of the engineer to act on behalf of the purchaser.

Clause 18(ii):

> The Contractor shall be responsible for ensuring that the positions, levels, and dimensions of the Works are correct according to the drawings, notwithstanding that he may have been assisted by the Engineer in setting out the said positions, levels, and dimensions.

This clause deals with the fairly common problem of incorrect setting out undertaken jointly with, or checked by, the engineer. Contractors are inclined to believe that the purchaser should accept some responsibility for the engineer's mistakes but to do so would have the effect of relieving the contractor of his own responsibility.

Clause 8(iii):

> All the Works shall be carried out under the direction and to the reasonable satisfaction of the Engineer.

Another way of saying that the contractor shall comply with the instructions of the engineer and that the engineer assesses whether or not specified standards have been achieved.

Engineer's representative

Clause 18(iv):

> The Engineer may from time to time delegate any of the powers, discretions, functions, and authorities vested in him and may at any time revoke any such delegation. Any such delegation or revocation shall be in writing signed by the Engineer and, in the case of a delegation, shall specify the power, discretions, functions, and authorities thereby delegated and the person or persons to whom the same are delegated. No such delegation or revocation shall have effect until a copy thereof has been delivered to the Contractor.

G/90 does not expressly require the engineer to appoint an engineer's representative – nor does it contain a definition of such a representative.

However clause 18(iv) clearly permits the engineer to delegate, without limits, to one or more persons. In this respect G/90 is more flexible than

MF/1 because the engineer can divide his functions between office based and site based staff.

Clerk of works

Clause 18(v):

> If a Clerk of the Works be appointed to watch the carrying out of the Contract, the Contractor shall afford him every reasonable facility for so doing, but the Clerk of the Works shall not be authorized to relieve the Contractor in any way of his duties or obligations under the Contract. Any written notice from the Clerk of the Works condemning any Plant or workmanship shall have the effect of a similar notice given by the Engineer under Clause 24 (Defects Prior to Taking-Over) except that the Contractor may appeal to the Engineer for his decision in the matter.

The clause does not state any procedure for the appointment of a clerk of works but good practice requires that it be in writing.

A clerk of works under G/90 is given both the power to watch and the power to condemn prior to taking-over. Even without delegated powers, therefore, a clerk of works under G/90 has more power than is commonly given to assistants.

Engineer's decisions

Clause 19:

> The Contractor shall proceed with the Works in accordance with decisions, instructions, and orders given by the Engineer in accordance with these Conditions, provided always that
>
> (a) if the Contractor shall, without undue delay after being given any decision, instruction, or order otherwise than in writing, require it to be confirmed in writing, such decision, instruction, or order shall not be effective until written confirmation thereof has been received by the Contractor, and
>
> (b) if the Contractor shall, by written notice to the Engineer within 14 days after receiving any decision, instruction, or order of the Engineer in writing or written confirmation thereof, intimate that he disputes or questions the decision, instruction, or order, giving his reasons for so doing, either party to the Contract shall be at liberty to refer the matter to arbitration pursuant to Clause 37 (Arbitration), but such an intimation shall not relieve the Contractor of his obligation to proceed with the Works in accordance with the decision, instructions, or order in respect of which the intimation has been given. The Contractor shall be at liberty in any such arbitration to rely on reasons additional to the reasons stated in the said intimation.

The effect of clause 19(a) is that the contractor must comply with written instructions of the engineer but need not comply with oral instructions.

But the clause is not written so as to invalidate oral instructions with which the contractor chooses to comply.

Clause 19(b), in conjunction with clause 37(1) (arbitration), acts as a condition precedent to arbitration in that disputes on the engineer's instructions, etc. can only be referred to arbitration if they are challenged by the contractor within 14 days after their receipt.

As with MF/1 only the contractor can lodge the initial challenge; but either party is then free to refer the matter to arbitration. Unlike MF/1 there is no specified procedure or timescale for the engineer's response and perhaps more surprisingly the clause does not even confirm that the engineer has the power to reconsider his position. But this can surely be implied.

Duties and powers

Schedules of the express duties and powers of the engineer in G/90 as given below show broad similarity with MF/1 although they are slightly less in number.

Express duties of the engineer under G/90

clause 4(i)	– to signify approval of drawings, etc within a reasonable period
clause 6(ii)	– to give consent as reasonable to sub-letting
clause 10(ii)	– to decide on variations which the contractor notifies will prevent or prejudice fulfilment of his obligations
clause 14(ii)	– to give 24 hours notice of intention to attend tests
clause 14(v)	– to confirm in writing when the plant has passed tests
clause 15(iv)(c)	– to certify in respect of delayed plant
clause 17(ii)	– to permit removal of contractor's equipment which is not required for the purposes of the works
clause 18(iii)	– to supervise the works
clause 24(b)	– to notify the contractor of alleged observed defects prior to taking-over
clause 25	– to grant such extensions of time as are appropriate and reasonable
clause 26	– to certify delay in completion
clause 28(i)	– to issue taking-over certificates for the works
clause 28(ii)	– to issue taking-over certificates for sections of the works
clause 28(iv)	– to issue taking-over certificates when the contractor is

	prevented from carrying out tests before completion by the purchaser, the engineer, or some other contractor employed by the purchaser
clause 30(ii)	– to notify the contractor of defects after taking-over
clause 31(iii)	– to issue interim payment certificates within 14 days of application
clause 31(vii)	– to issue a final certificate within 28 days of application
clause 32(i)	– to deal with provisional sums
clause 32(ii)	– to deal with PC items
clause 36(ii)	– to direct on dimensions in the event of metrication problems.

Express powers of the engineer under G/90

clause 4(i)	– to request the submission of drawings, etc for approval
clause 4(iii)	– to inspect drawings of the premises of the contractor
clause 8(i)	– to issue instructions on the manner of execution of the works
clause 8(iii)	– to issue instructions to vary the programme
clause 10(i)	– to issue variations and omissions to the works
clause 14(i)	– to inspect, examine and test plant on the contractor's premises
clause 14(vi)	– to reject plant not in accordance with the contract
clause 15(i)	– to authorize delivery of plant to site
clause 15(vi)	– to require examination and making good of delayed plant
clause 16(v)	– to direct times of work
clause 17(iv)	– to withhold payment on plant which has not been set aside and marked
clause 18(i)	– to give instructions and orders
clause 18(iv)	– to delegate powers and functions
clause 18(v)	– to consider appeals on decisions of the clerk of works
clause 20(ii)	– to object to the presence on site of the contractor's employees on the grounds of misconduct, etc
clause 24(a)	– to reject work considered to be defective prior to taking-over
clause 27(i)	– to fix the dates of tests before completion
clause 27(iii)	– to fix the time for delayed tests before completion
clause 27(iii)	– to make tests before completion in the event of the contractor failing to do so
clause 27(v)	– to require tests before completion which have failed to be repeated
clause 27(vi)	– to extend time for the passing of performance tests

clause 28(vi)   – to require examination of the works prior to the
                    carrying out of delayed tests on completion
clause 29   – to suspend the progress of the works
clause 31(xi)   – to correct certificates
clause 32(i)   – to instruct on the use on provisional sums
clause 32(ii)   – to instruct on the use of prime cost items.

Chapter 9

Inspection and testing

9.1 Introduction

This chapter examines the provisions in the model forms for the inspection and testing of plant prior to taking-over. Also considered are the circumstances in which tests, other than performance tests, can be required during the defects liability period.

Performance tests are covered in Chapter 13.

Details of tests

The provisions for testing in the model forms are procedural only. They are not concerned with the details of tests. These need to be set out with clarity and precision in the specification.

Sample specifications for tests before delivery and tests on completion are given at the end of this chapter.

Purpose of tests

Testing in plant contracts can be undertaken at three stages – each with its own purpose:

- tests before delivery – to ensure that the plant has been manufactured to specification and that each item performs to specification and/or manufacturers' guarantee standards;
- tests on completion – to ensure that the erected plant operates properly and can be put into use;
- tests after completion (performance tests) – to ascertain if the performance of the plant under working conditions is up to required standards.

Importance of tests

All tests are important but the significance of passing tests and the consequences of failing tests differs for each of the categories of tests:

- tests before delivery – passing the tests entitles the contractor to deliver and usually to some payment. Failing the tests prejudices delivery and may lead to rejection of that part of the plant which is defective;
- tests on completion – passing the tests leads to taking-over; transfer of responsibility for care of the works; commencement of the defects liability period; and payment. Failing the tests may cause delay; liability for damages for late completion; the need for repetition of tests; the possibility of rejection of the plant;
- performance tests – passing the tests leads to acceptance of the plant and fulfilment of the contractor's obligations. Failing the tests may lead to modifications of the plant; liability for damages for low performance; rejection of the plant.

Tests on completion

The tests of greatest contractual significance are the tests on completion. It is on successful conclusion of these tests that the engineer certifies to the effect that the plant is fit for use and the purchaser takes control. Providing the contractor then meets his obligations in respect of defects and the performance tests are satisfactory the contractor will at the end of the defects liability period automatically become entitled to a final certificate confirming that he has fulfilled his obligations under the contract.

Thus the tests on completion have definitive status in relation to both time and quality in the contract subject only to considerations of defects and performance.

9.2 *Inspection and testing in MF/1*

The relevant provisions of MF/1 are:

- clause 1.1.v   – definition of tests on completion
- clauses 23.1 to 23.5 – inspection and testing of plant before delivery
- clauses 28.1 to 28.5 – tests on completion
- clause 31.1   – interference with tests
- clause 31.2   – tests during the defects liability period
- clause 36.7   – further tests.

Definition of tests on completion

Clause 1.1.v

'Tests on completion' means the tests specified in the Contract (or otherwise agreed by the Purchaser and the Contractor) which are to be made by the

Contractor upon completion of erection and/or installation before the Works are taken over by the Purchaser.

Although this definition permits tests on completion to be 'otherwise agreed' it is strongly recommended that tests on completion should be specified in the contract wherever practicable.

The definition although simple makes two important points:

- taking-over follows the tests on completion;
- the contractor should know what is required by way of testing in order to obtain taking-over.

Inspection and testing of plant before delivery

Clause 23.1 – Engineer's entitlement to test

> The Engineer shall be entitled at all reasonable times during manufacture to inspect, examine and test on the Contractor's premises the materials and workmanship and performance of all Plant to be supplied under the Contract. If part of the Plant is being manufactured on other premises the Contractor shall obtain for the Engineer permission to inspect, examine and test as if the Plant were being manufactured on the Contractor's premises. Such inspection, examination or testing shall not release the Contractor from any obligation under the Contract.

The power of the engineer to 'test' himself needs to be treated with caution. Any negligence by the engineer could lead to his own and/or, the purchaser's liability. See the 'Purchaser's Risks' in clause 45.1. The engineer should do no more than witness tests unless personal involvement is essential.

The provision for examination and testing on the premises of sub-contractors and suppliers should encourage the contractor to include a 'back to back' clause in his order. There are no ordinary legal rights of entry or observation.

The application of the last sentence of the clause is not wholly clear. If it applies only to the preceding sentence it means that tests, etc on sub-contractors' or suppliers' premises do not release the contractor from any obligation. If it applies to the whole clause it means that no off-site tests by the engineer release the contractor from any obligation. The latter interpretation is probably the best since testing, etc under this clause is preliminary to testing 'provided in the Contract' as covered by clause 23.2.

Clause 23.2 – dates for test and inspection

> The Contractor shall agree with the Engineer the date on and the place at which

any Plant will be ready for testing or inspection as provided in the Contract. The Engineer shall give the Contractor 24 hours' notice of his intention to attend the test or inspection. If the Engineer shall not attend at the place so named on the date agreed, the Contractor may proceed with the test or inspection which shall be deemed to have been made in the Engineer's presence. The Contractor shall forthwith forward to the Engineer duly certified copies of the results of such test or inspection.

This clause applies only to testing or inspection as provided in the contract – that is, to those tests detailed in the specification.

The attendance of the engineer is not mandatory and failure by the engineer to attend is not a default of any consequence, if at all. However, non-attendance by the engineer does entitle the contractor to certify his own results – an arrangement which might not be to the liking of the purchaser.

Clause 23.3 – services for tests and inspections

Where the Contract provides for tests or inspections on the premises of the Contractor or of any Sub-Contractor, the Contractor shall provide free of charge such assistance, labour, materials, electricity, fuel, stores, apparatus and instruments as may be requisite and as may be reasonably demanded to carry out such test or inspection.

The obligations imposed on the contractor in this clause would seem to go without saying. But the clause does emphasize the difference between obligations for off-site tests and on-site tests – where under clause 11.7 the purchaser is responsible for certain supplies.

Clause 23.4 – certificate of testing or inspection

When the Engineer is satisfied that any Plant has passed the test or inspection referred to in this Clause he shall forthwith issue to the Contractor a certificate to that effect.

The purpose of a certificate issued under this clause is not stated. Its most obvious use is in confirming that plant is suitable for delivery. In some cases it may also be linked to interim payments – but that is an arrangement not expressly covered in MF/1.

The contractor can probably argue for a deemed certificate if the engineer fails to attend tests and subsequently declines to issue a certificate.

Clause 23.5 – failure on tests or inspection

If after inspecting, examining or testing any Plant the Engineer shall decide that such Plant or any part thereof is defective or not in accordance with the Contract, he may reject the said Plant or part thereof by giving to the Contractor within 14 days notice of such rejection, stating therein the grounds upon which the said decision is based. Following any such rejection, the Contractor shall make good or otherwise repair or replace the rejected Plant and resubmit the same for test or inspection in accordance with this Clause and all expenses reasonably incurred by the Purchaser in consequence of such re-testing or inspection and the Engineer's attendance shall be deducted from the Contract Price.

This clause is intended to operate as follows:

- inspecting, examining or testing
  ↓
- rejection by the engineer
  ↓
- notice with reasons to be given by the engineer within 14 days
  ↓
- repair, replacement or resubmission for testing, etc
  ↓
- retesting
  ↓
- purchaser's costs of retesting and the engineer's attendance to be deducted from the contract price.

It is not obvious what costs the purchaser can incur other than the costs of the engineer's attendance since by clause 23.3 the contractor is made responsible for the costs of tests and inspections.

Tests on completion

Clause 28.1 – notice of tests

The Contractor shall give to the Engineer 21 days' notice of the date after which he will be ready to make the Tests on Completion. Unless otherwise agreed the Tests on Completion shall take place within 10 days after the said date on such day or days as the Engineer shall notify the Contractor.

This clause means that the contractor is entitled to make the tests on completion within 31 days of giving notice that he will be ready – 21 days notice plus 10 days for agreement of the date of testing. A common problem is that the contractor, having given notice of being ready, does not achieve his expected progress. In such circumstances the original notice, it is suggested, falls and re-serving is necessary.

Clause 28.2 – time for tests

> If the Engineer fails to appoint a time after having been asked so to do or to attend at any time or place duly appointed for making the Tests on Completion, the Contractor shall be entitled to proceed in his absence and the Tests on Completion shall be deemed to have been made in the presence of the Engineer. The Contractor shall forward to the Engineer duly certified copies of the results of the Tests on Completion.

This clause reveals a modest flaw in the preceding clause in that the time allowed to the engineer for fixing the date of tests after being given notice is not specified.

It cannot be intended that the engineer has 31 days to respond to the contractor's notice consequently it is difficult to fix precisely when the phrase 'If the Engineer fails to appoint a time after having been asked so to do' comes into effect.

A practical solution is for the contractor when giving notice to specify the intended date of the tests with the proviso that the intended date is flexible within the specified 10 day period should the engineer prefer another date.

However, having regard to the importance of the tests on completion, it would be unwise of the engineer not to appoint a time and more so not to attend the tests.

Clause 28.3 – delayed tests

> If the Tests on Completion are being unduly delayed by the Contractor, the Engineer may, by notice, call upon the Contractor to make them within 21 days from the receipt of the said notice. The Contractor shall make the Tests on Completion on such days within the said 21 days as the Contractor may fix and of which he shall give notice to the Engineer. If the Contractor fails to make the Tests on Completion within the time aforesaid the Purchaser or the Engineer may proceed therewith at the risk and expense of the Contractor and the Cost thereof shall be deducted from the Contract Price. If the Contractor shall establish that the Tests on Completion were not being unduly delayed, the Tests so made shall be at the risk and expense of the Purchaser.

This is a difficult and a dangerous clause to apply in practice.

The problem starts with the phrase 'being unduly delayed'. This may mean either that the contractor has not proceeded with tests after giving notice or that the time for completion has passed and no notice of readiness has been given. The better interpretation is probably that the time for completion has passed.

The next problem is the phrase 'the Purchaser or the Engineer may proceed therewith'. This apparently means – since there is a reference to the cost thereof – proceed with the tests. But given the express mention of the purchaser a more appropriate practical meaning is – proceed with using the works.

As to the phrase 'the risk and expense of the Contractor', expense is one thing but the transference of risk another. If by negligence the works are damaged by the purchaser or the engineer it is unlikely that the provisions of clause 45.1 (purchaser's risks) cease to apply.

The final sentence of the clause contains more difficulties. How is the contractor to establish that the tests on completion were not being unduly delayed? Obtaining an extension of time for completion is one possibility – but, that apart, there is no contractual mechanism for dealing with this situation. And finally, if the contractor does establish that the tests were not being unduly delayed what is the intention if the tests fail – are they still to be at the expense of the purchaser?

This latter point reveals the fundamental problem in the clause. If the contractor delays in testing because he does not believe the works to be ready and the purchaser or engineer take it upon themselves to test – and damage the works in the process, they will effectively prove the contractor's case – that the tests are not being unduly delayed.

Clause 28.4 – repeat tests

> If any part of the Works fails to pass the Tests on Completion they shall be repeated within a reasonable time upon the same terms and conditions. All Costs which the Purchaser may incur in the repetition of the Tests on Completion shall be deducted from the Contract Price. The provisions of this Sub-Clause shall also apply to any tests carried out under Sub-Clause 36.7 (Further Tests).

The opening words of the following clause highlight a deficiency here. Clause 28.5 commences 'If the Works or any Section' and presumably clause 28.4 is also intended to apply to the whole of the works as well as to parts.

Unlike off-site tests the purchaser's costs of a repeat testing extend well beyond the attendance of the engineer. See clause 11.7 on the purchaser's obligation to provide things necessary for tests on site.

Clause 28.5 – consequences of failure to pass tests on completion

> If the Works or any Section fails to pass the Tests on Completion (including any repetition thereof) the Contractor shall take whatever steps may be necessary to

enable the Works or the Section to pass the Tests on Completion and shall thereafter repeat them, unless any time limit specified in the Contract for the passing thereof shall have expired, in which case the Engineer shall be entitled to reject the Works or the Section and to proceed in accordance with Clause 49 (Contractor's Default).

To make this clause work it is essential that a time for passing tests on completion is stated in the special conditions. Otherwise the right of rejection under clause 49 cannot be established under this clause – some other grounds would have to be used.

But in any event this clause needs to be considered in conjunction with clause 30.1 (use before taking-over) which entitles the purchaser to use the works, if they are capable of being used, when a taking-over certificate is not issued within one month of the date for completion. There are many situations in practice where the purchaser has little option but to put the works into use even though the contractor is experiencing difficulty passing the tests on completion. Rejection, therefore, is not the only nor necessarily the appropriate course of action.

Clause 28.5 does not, however, offer any solution other than continued repetition of the tests or rejection. And in so far that continued repetition leads eventually to prolonged delay as considered in clause 34.2 – that also has only one positive ending – termination.

The implication in the last sentence of clause 28.5 that it is the engineer who 'proceeds' with termination in accordance with clause 49 is not intended and it is likely to be eliminated in the 1995 Revision 3 to MF/1 by words to the effect that the purchaser shall be entitled to proceed with termination.

Interference with tests

Clause 31.1 – interference with tests

> If by reason of any act or omission of the Purchaser, the Engineer or some other contractor employed by the Purchaser, the Contractor shall be prevented from carrying out the Tests on Completion in accordance with Clause 28 (Tests on Completion) then, unless in the meantime the Works have been proved not to be substantially in accordance with the Contract, the Purchaser shall be deemed to have taken over the Works and the Engineer shall, upon the application of the Contractor, issue a Taking-Over Certificate accordingly.

This clause deals positively with prevention in stating that deemed taking-over occurs and that the engineer shall issue a taking-over certificate. It puts beyond doubt matters which are frequently left open for argument in other contracts. For the contractor it is a good protective provision but for the purchaser it can be a cause for concern.

The problem is that the clause comes into effect without giving the purchaser a reasonable period to rectify the prevention for which he is responsible. Under clause 28.1 tests are to take place within a 10 day period after expiration of the contractor's notice of readiness to test and once the 10 day period has elapsed the contractor can claim deemed taking-over.

There may be some relief for the purchaser in the phrase 'unless in the meantime the Works have been proved not to be substantially in accordance with the Contract'. This suggests that there is an indeterminate period ('the meantime') between the due date for testing and the date of deemed taking-over when proof can be sought as to whether the works are substantially in accordance with the contract. But this may be putting a literal interpretation on the clause which is not intended. A more practical meaning would be that deemed taking-over does not occur, notwithstanding prevention, if it can be shown that the works were not capable of passing the tests on completion at the time they were due.

Tests during the defects liability period

Clause 31.2 – tests during defects liability period

> In any case where a Taking-Over Certificate has been issued under Sub-Clause 31.1 (Interference with Tests) the Contractor shall be under an obligation to carry out the Tests on Completion during the Defects Liability Period as and when required by 14 days' notice from the Engineer. Such allowances shall be made from the results required to be attained in the Tests on Completion as may be reasonable having regard to any use of the Works by the Purchaser prior to the Tests on Completion and to any deterioration therein which may have occurred since the issue of the Taking-Over Certificate in respect thereof. The additional Costs incurred by the Contractor in making the Tests on Completion in accordance with this Sub-Clause shall be certified by the Engineer and added to the Contract Price.

The requirement in this clause for the contractor to make the delayed tests on completion as and when required during the defects liability period seems reasonable enough. But the question is – what purpose do such tests serve? What are the effects of success? And what are the consequences of failure?

Presumably if the tests are successful nothing changes. The contractor already has his taking-over certificate and the defects liability has already commenced.

If the tests fail the position is potentially complex. There is no provision in MF/1, as there is in the IChemE model forms for revocation of a certificate when delayed tests fail.

Repeat tests could be called for under clause 28.4 – providing that the defects liability period has not elapsed. Clause 28.5 (consequences of failure to pass tests on completion) could perhaps be applied. But this clause contemplates only repetition or rejection; and rejection when the works are already in use is neither a sound practical nor legal solution.

Even the question of whether the defects liability period is extended following the failure of tests is open to argument. Clause 36.3 (notice of defects) and clause 36.4 (extension of defects liability) may be relevant in some cases but these clauses do not have the effect which the purchaser would no doubt wish to see of the defects liability period being re-fixed to run from eventual successful passing of the tests.

Another aspect of clause 31.2 which the purchaser might find unacceptable in the way it is worded is the requirement to pay the contractor the additional costs of making the tests on completion during the defects liability period. The intention is almost certainly that the purchaser should pay only for successful tests but that is not specified. It does say in clause 28.4 (repeat tests) that the costs of repeat tests shall be deducted from the contract price but that clause may not apply to repeat tests under clause 31.2 since it is only expressly extended in its application to clause 36.7 (further tests).

Further tests

Clause 36.7 – further tests

> If the repairs or replacements are of such a character as may affect the operation of the Works or any part thereof, the Purchaser or the Engineer may within one month after such repair or replacement give to the Contractor notice requiring that further Tests on Completion or Performance Tests be made, in which case such Tests shall be carried out as provided in Clauses 28 (Tests on Completion) or 35 (Performance Tests) as the case may be.

This clause applies only when the works show evidence of defects or damage during the defects liability period.

For performance tests the consequences of failure are as set out in clause 35.8 but for tests on completion the consequences of failure are again uncertain. For further comment see Chapter 14 on liability for defects.

The costs of making further tests will normally fall on the contractor.

9.3 Inspection and testing in MF/2

MF/2 contains significantly fewer provisions on inspection and testing than MF/1. This is because there is no place in MF/2 for tests on completion and related provisions on interference with such tests.

Accordingly the only provisions in MF/2 covering inspection and testing are:

- clause 1.1.t   – definition of tests
- clauses 17.1 to 17.5 – inspection and testing of plant before delivery
- clause 25.6   – further tests.

Definition of tests

Clause 1.1.t

'Tests' means the tests specified in the Contract (or otherwise agreed by the Purchaser and the Contractor) which are to be made by the Contractor upon completion of manufacture of the Plant before Delivery.

MF/2 does not have any tests before completion – it has only tests before delivery – which by the above definition are termed simply 'tests'.

These tests serve a similar contractual purpose to tests on completion in MF/1 by marking the point at which the contractor has fulfilled his obligations in respect of manufacture (manufacture and erection in MF/1). But whereas in MF/1 – taking-over follows the tests on completion thereby indicating fulfilment of the contractor's primary obligations, in MF/2 delivery is still to follow the tests and it is delivery which marks fulfilment of the contractor's primary obligations.

Inspection and testing of plant before delivery

Clauses 17.1 to 17.5 of MF/2 correspond almost word for word with clauses 23.1 to 23.5 of MF/1. The only notable difference is in clause 17.3 (services for tests or inspection) where MF/2 omits the MF/1 reference to tests on the premises of sub-contractors.

Further tests

Clause 25.6 of MF/2 is similar in purpose to clause 36.7 of MF/1

If the repairs or replacements are of such a character as may affect the operation of the Plant or any part thereof, the Purchaser or the Engineer may notify the Contractor that the Purchaser will repeat such tests thereof as were carried out prior to Delivery. The costs reasonably incurred by the Purchaser in such tests so notified shall be deducted from the Contract Price. The Contractor may attend such tests on reasonable notice to the Purchaser.

For practical reasons in MF/2 it is the purchaser and not the contractor (as in MF/1) who carries out the further tests – hence the provision that the contractor may attend. The contractual consequences of failure are as difficult to predict as in MF/1.

9.4 Inspection and testing in MF/3

MF/3 provides for inspection and testing before delivery in similar manner to MF/2. The relevant clauses are:

- clause 4.1 – inspection and testing before delivery
- clause 4.2 – prescribed tests
- clause 4.3 – failure on testing.

Inspection and testing before delivery

Clause 4.1

> Before delivering any goods the Vendor shall inspect and test the same for compliance with the Contract and, if so requested, shall supply to the Purchaser a certificate of the results of the test.

This clause imposes a general obligation on the vendor which is expressed independently of prescribed tests. It is of direct concern to the purchaser only in so far that he requires test certificates.

Prescribed tests

Clause 4.2

> Where the Contract provides that the goods shall pass any prescribed tests or shall give a specified performance they shall be tested by the Vendor before delivery for compliance with the prescribed tests or for performance or for both as the case may be, the Vendor providing free of charge what may be requisite for the purpose. The Vendor shall give the Purchaser seven days' notice in writing of the date on and the place at which any of the goods will be ready for testing as provided in this sub-clause. If the Purchaser shall fail to give the Vendor 24 hours' notice, or shall fail to attend on the day he has appointed, the Vendor may proceed with the tests, which shall be deemed to have been made in the Purchaser's presence. The Vendor shall forthwith forward to the Purchaser a certificate of the results of the tests.

This clause deals with specific obligations to tests. The purchaser is entitled to attend such tests but the vendor may proceed if the purchaser fails to attend.

Failure on testing

Clause 4.3

> If on a test made pursuant to Sub-Clause 4.2 (Prescribed Tests) the goods or any part thereof shall fail to pass the prescribed tests or to give the specified performance such goods or part thereof shall, if the Vendor so desires, be tested again or the Vendor may submit for test other goods in their place. If the goods or the said other goods shall fail to pass the test or to give the specified performance, the Purchaser shall be entitled by notice in writing to reject the goods or such part thereof as shall have failed as aforesaid.

The status of prescribed tests in MF/3 is clearly revealed in this clause. Goods which fail to pass the tests may be rejected. However, first the vendor is entitled to re-test once but, from the wording of the clause, if the tests then fail the purchaser need not wait for further re-tests before rejecting the goods.

9.5 Inspection and testing in G/90

G/90 and MF/1 have similar provisions on inspection and testing with both being developed from the old MF/A.

In G/90 the terminology is 'tests before completion' rather than 'tests on completion' but the two are effectively the same.

The relevant provisions of G/90 are to be found in:

- clause 1   – definition of tests before completion
- clause 14 – inspection, testing and rejection
- clause 27 – tests before completion
- clause 28 – interference with tests
- clause 30 – further tests.

Definition of tests

Clause 1

> 'Tests before Completion' shall mean such tests to be made by the Contractor on completion of erection as are provided for in the Contract or otherwise agreed between the Purchaser and the Contractor.

This is not as positive a definition as in MF/1 in that it avoids the point that the tests before completion are tests precedent to taking-over. But that point, nevertheless, is covered in clause 27(vi).

Inspection, testing and rejection

The provisions of clause 14 of G/90 and clause 23 of MF/1 compare as follows:

- engineer's entitlement to test
  - clause 14(i) of G/90; clause 23.1 of MF/1
  - identical
- dates for test and inspection
  - clause 14(ii) of G/90; clause 23.2 of MF/1
  - virtually identical, but G/90 clause refers only to tests
- services for tests and inspection (on the contractor's premises)
  - clause 14(iii) of G/90; clause 23.3 of MF/1
  - virtually identical, but G/90 includes the phrase 'except where otherwise specified' and again refers only to tests
- services for tests and inspection on site
  - clause 14(iv) of G/90; clause 11.7 of MF/1
  - virtually identical
- certificate of testing or inspection
  - clause 14(v) of G/90; clause 23.4 of MF/1
  - G/90 refers only to tests and instead of a certificate simply requires the engineer to notify the contractor in writing when he is satisfied the plant has passed its tests
- failure on tests or inspections
  - clauses 14(vi) and 14(vii) of G/90; clause 23.5 of MF/1
  - in G/90 clauses 14(vi) and 14(vii) have to be read with clause 27(v) on failure to pass tests to match the provisions of clause 23.5 of MF/1. Taken together the provisions are nearly identical except that the G/90 clause requires the engineer to give the contractor notice of failure 'within a reasonable time' whereas MF/1 requires notice within 14 days.

Tests before completion

Clause 27 of G/90 compares with the corresponding clause 28 of MF/1 as follows:

- Notice of tests
  - clause 27(i) of G/90; clause 28.1 of MF/1
  - identical

- time for tests
  - clause 27(ii) of G/90; clause 28.2 of MF/1
  - the same except that the G/90 clause does not require the contractor to produce certified copies of results of tests made in the engineer's absence
- delayed tests
  - clause 27(iii) of G/90; clause 28.3 of MF/1
  - similar except that G/90 allows the contractor 10 days to proceed with tests against 21 days in MF/1
- services for tests
  - clause 27(iv) of G/90; clause 11.7 of MF/1
  - virtually identical
- repeat tests
  - clause 27(v) of G/90; clause 28.4 of MF/1
  - virtually identical
- status of tests
  - clause 27(vi) of G/90; clause 1.1.v of MF/1
  - both clauses make completion dependent on passing the tests.

Interference with tests

The provisions of clause 28(iv) of G/90 are effectively the same as those in clause 31.1 and clause 31.2 of MF/1 with the minor exception that in G/90 the engineer is obliged to issue a taking-over certificate following interference with tests whereas in MF/1 the engineer need only issue a certificate on the application of the contractor.

Further tests

Clause 30(v) of G/90 matches the provisions in clause 36.7 of MF/1 regarding further tests following repairs, renewals or replacements.

9.6 Sample specification – tests before delivery

Tests for various items of Plant shall be carried out, as follows:

Pumps

All pumps shall be tested individually in accordance with BS5316. Tests shall be to Part 2 (Class B). Site conditions shall be simulated as near as possible including the NPSH condition. Pumps shall be tested with their own prime movers. Where it is impractical to include the full length of the connecting shaft, the Contractor shall state the allowances to be made for the losses incurred by its omission and shall demonstrate the accuracy of the allowances to the satisfaction of the Engineer.

Each pump shall be tested at its guaranteed duty point and over its full working range from its closed valve condition to 20% in excess of the specified quantity as minimum head. Tests shall provide information for performance curves to be drawn for: head/quantity, efficiency/quantity, power absorbed/quantity and net positive suction head/quantity.

Pump casings shall be subjected to a pressure test at 1.5 times the pressure obtained with the delivery valve closed. The positive suction head shall be taken into account in determining this pressure.

In addition to confirming the hydraulic performance of the pumpset as specified, the test shall demonstrate that vibration is within the specified limits and that the mechanical performance is satisfactory.

Pipework and valves

Pipes, pipe fittings, and valve bodies shall be pressure tested (with closed ends) to 1.5 times the rated pressure in accordance with the appropriate standards.

Butterfly valves with rubber seats shall be tests to the maximum differential pressure, at which pressure they shall be drop-tight.

Butterfly valves with metal seats shall be tested to the maximum differential pressure and the permissible leakage rate shall not exceed 0.10 litres/hour/100 mm of the diameter with on-seat pressure.

Castings

Castings subject to hydraulic pressure shall be tested to 1.5 times the maximum working pressure for a period of at least one hour.

Cranes and lifting equipment

Cranes and lifting equipment shall be completely assembled and tested for all operations in accordance with BS466.

Cranes and lifting equipment shall also be tested after erection on site. They shall not be used until the tests have been completed satisfactorily and a test certificate issued. Slings and tackle shall be tested and must not be used until test certificates have been issued.

Structural sections and materials

For structural sections and materials used in the construction of the Plant, the Contractor shall provide manufacturer's certificates of tests, proof sheets, mill sheets, etc showing that the materials have been tested to the appropriate British Standard.

Compressors and blowers

Tests shall be performed in accordance with the relevant British Standard. All compressors and blowers shall be tested with their ancillaries to confirm design performance particularly in respect of flow and pressure. The test shall demonstrate that vibration and noise are within the specified limits and that the pressure relief valve operates correctly.

Electric motors

Electric motor tests shall be performed in accordance with the requirements of BS4999 as applicable. The tests shall obtain the overall efficiency and other figures in accordance with the guarantees given in Technical Data Sheets.

Switchgear and motor control centres

The whole of the switchgear and controlgear shall be witness tested as integral units for a complete sequence of operation and as laid down in EN 60439 and based on the completeness of the circuits in the final manufactured form within the manufacturer's works.

The following tests shall be performed:

- Primary injection tests to ensure correct rates and polarity of CTs and VTs and of the current operated protection relays and direct acting coils, over their full range of settings;
- Balance earth fault stability test by primary current injection. Care must be taken to reproduce accurately the burdens of interconnecting cables. A further test to ensure correct polarity must be made after assembly;
- Tests on auxiliary relays at normal operating voltages by operation of associated remote relays;
- Correct operation of sequencing and control circuits at normal operating voltages by operation of local control switches, and simulation of operation from remote control positions.

Circuit breakers

All circuit breakers shall be subject to the following tests:

- Routine tests including HV pressure test, milli-volt drop tests and mechanical tests;
- To ensure the operation of the DC closing coil and satisfactory closing of the circuit breaker with the voltage of the coil down to 80% of its rated voltage, and that mal-operation does not occur with a voltage on the coil of 120% of its rated voltage;
- Interchangeability of withdrawable identically equipped circuit breakers, and checking of all mechanical and electrical inter-locks.

Type test figures for heat runs performed on identical panel types shall be made available.

Transformers

Transformers shall be works routine tested, which shall also include the following:

- Measurement of winding resistance
- Ratio polarity and phase relationship
- Impedance voltage
- Load losses
- No load losses and no load current
- Insulation resistance
- Induced overvoltage withstand
- Separate source voltage withstand.

Type test certificates shall be provided for the following:

- Impulse voltage withstand
- Temperature rise.

Cables

All HV cables and armoured cables shall be subject to routine tests in accordance with the relevant British Standard specification. Test certificates shall be provided against each drum and/or cable length.

The tests performed on every cable length and/or drum at manufacturer's premises shall include:

- High voltage DC insulation pressure test, between cores, each core to earth, metallic sheath or armour as applicable
- Insulation resistance test
- Core continuity and identification
- Conductor resistance test.

Pressure switches and gauges

All pressure switches and vacuum and pressure gauges shall be subject to routine tests in accordance with the relevant British Standard specification.

9.7 Sample specification – Tests on completion

When the Plant or portion of Plant has been erected and is ready to be put into service, the Contractor shall carry out the Tests on Completion. These tests shall comprise the pre-commissioning tests and on-load measurements.

At least three months before Tests on Completion, the Contractor shall submit to the Engineer his detailed proposals for testing prior to use of equipment, energization and on-load operation. The proposals shall state values of test parameters, and make reference to standards and manufacturer's literature. The proposed format of test sheets shall be submitted at the same time. A separate sheet shall be used for each test. The testing shall not commence until the proposals and test sheets have been approved in writing by the Engineer.

At least one week before the Contractor intends to start testing, he shall submit a provisional testing programme to the Engineer. The programme shall be updated and resubmitted weekly during the testing. The programme shall indicate which tests the Contractor intends to carry out on a daily basis.

All measuring instruments, indicators, test sheets, test equipment, supervision and labour for Tests on Completion shall be provided by the Contractor at his own cost and shall be included in the prices for the erection of the Plant.

In general, the Tests on Completion shall consist of:

- an inspection to check assembly of the Plant;
- a check of conformity with the Specification;
- functional tests to ensure that the Plant works as the manufacturers intend;
- testing or checking of site installations, e.g. earthing values, cabling, pipework fluid levels, insulation, relay settings, hydraulic and leak tests;

- a four-hour performance test on every motor to prove:
  - correct functioning
  - absence of fluid leaks
  - correct bearing temperatures
  - absence of undue vibration or noise.

During this test a check on the performance of the equipment shall be made, as far as site facilities will allow. The site performance shall be compared with the official factory tests to identify any constraints on performance due to site conditions. The initial charges of oil, grease, electrolyte and similar materials necessary for the tests on completion shall be provided by the Contractor.

Where applicable, additional tests shall be performed as stated in other parts of the specification.

Chapter 10

Delivery and suspension

10.1 Introduction

Plant contracts rarely stand alone and usually the engineer or the purchaser has the task of co-ordinating the work of plant contracts with other contracts.

For this reason provisions for delaying delivery or suspending the progress of the works are far more likely to be put into effect in plant contracts than in process or construction contracts.

Powers to suspend

It is generally held that without an express provision for suspension neither party can suspend, nor can the engineer order suspension. To suspend without express provision amounts either to breach of contract or reliance on an implied term. The difficulty with the implied term is that an assumption has to be made to justify the action but the courts will only agree if it is necessary to give business efficacy of the contract.

In the case of *Canterbury Pipe Lines Ltd* v. *Christchurch Drainage Board* (1979), in a dispute over amounts certified by the engineer on a drainage contract, the contractor suspended work which led to the employer taking the work out of the contractor's hands. It was held that the contractor had no right to suspend, and in commenting but not ruling on the general right of the contractor to suspend Mr Justice Cooke said this:

'We are against recognizing such a right when the architect or engineer has declined to issue the certificate stipulated for by the contract as a condition precedent to the employer's duty to pay. It would disrupt the scheme of these contracts. It could encourage contractors to take the law into their own hands. They might stop work which in the public interest needs to be done promptly. In such cases, if the contractor cannot or does not wish to rescind and cannot prove impossibility or its equivalent, he will be left with whatever remedies regarding the recovery of progress payments may be available to him under the contract. As already indicated we do not exclude the view that a case of impossibility or its equivalent might be made out by proving that for want of money the contractor could not carry on or could not reasonably be expected to do so.'

So much for the position of the contractor taking action into his own hands. As for the position of the engineer or the purchaser suspending work without an express provision that would almost certainly amount to prevention and a breach of contract.

All four model forms for plant recognize this point and make express provision for the engineer and/or purchaser to delay delivery of the plant or suspend the progress of the works.

Reasons for suspensions

Suspensions and delayed deliveries arise or are instructed for five broad categories of reasons:

- breaks in performance/progress specified in the contract;
- defaults in performance for which the contractor is responsible
- other matters for which the contractor is contractually responsible;
- delays caused by the engineer or the purchaser;
- delays caused by others for whom the purchaser is contractually responsible.

Remedies for suspensions

As a general rule, the contractor is expected to allow in his price for suspensions specified in the contract. And as a general rule, the contractor will have no remedy for suspensions for which he is responsible. But see the comments on MF/1 which follow.

For suspensions for which the purchaser is responsible a scale of remedies is usually provided to the contractor as follows:

- entitlement to costs arising;
- entitlement to interim payment after a specified time;
- entitlement to terminate the contract after a further specified time.

A notable difference between MF/1 and G/90 is that whereas MF/1 includes all three remedies, G/90 stops short of expressly providing for termination by the contractor.

10.2 Delivery and suspension in MF/1

Clause 24.1 of MF/1 covers delivery of the plant to site. Clauses 25.1 to

25.6 cover suspension of the works which in MF/1 expressly includes suspension of deliveries.

Clause 24.1 – delivery

Clause 24.1 contains three clearly stated provisions, namely:

- the contractor must apply to the engineer for permission to deliver plant or contractor's equipment to site;
- no plant or contractor's equipment may be delivered to site without the engineer's written permission;
- the contractor is responsible for reception and unloading of all plant and contractor's equipment.

Failure by the engineer to give permission is dealt with in clause 25.1.

Clause 25.1 – instructions to suspend

Clause 25.1 commences with a general provision empowering the engineer to suspend the progress of the works. In the 1995 Revision 3 it is expected that, for the avoidance of doubt, the power to suspend will expressly relate to the whole or any part of the works.

The phrase used in clause 25.1 is 'at any time' which indicates that suspension can apply to off-site as well as on-site activities. The engineer is not expressly required to give reasons for any suspension but later provisions in MF/1 dealing with costs make it clear that if there is to be any disallowance of the contractor's costs then the reasons have to be declared.

Clause 25.1 concludes with two further general provisions:

- that the contractor shall store, preserve and protect the works during suspension and insure as required by the engineer;
- that, unless otherwise instructed, the contractor shall maintain his resources (staff, labour, equipment) on or near the site ready to proceed when so instructed.

This latter provision applies only to suspensions affecting the progress of the works on the site and it runs counter to the usual rule that the contractor has a duty to mitigate his costs. What clause 25.1 requires is that the contractor's organization on site should remain at full strength unless the engineer instructs otherwise. Clearly this is a matter which any engineer issuing a suspension must attend to.

The provision does not state the position in respect of suspensions affecting work off-site. It is suggested that in such cases the duty to mitigate applies.

Clauses 25.1(a) and (b) – deemed suspension

The central part of clause 25.1 deals with deemed suspensions when there is prevention for which the purchaser is responsible.

That responsibility is detailed in the clause as:

- any delay or failure on the part of the engineer;
- failure by the engineer to give permission for delivery;
- any cause for which the purchaser is responsible;
- any cause for which some other contractor employed by the purchaser is responsible.

The prevention to which clause 25.1 applies is detailed as:

- clause 25.1(a) – delivery to site of plant that is ready at the time specified in the programme – or if no time is specified at a time appropriate having regard to the time for completion;
- clause 25.1(b) – erecting any plant that has been delivered to site.

In the event that there is such prevention and the purchaser is responsible the engineer is deemed to have given instructions to suspend to the extent that progress is dependent on delivery or erection of the affected plant.

These provisions have enormous potential to the contractor in claims for additional payment. Their scope is wide enough to extend to many of the common delays and interruptions suffered by contractors, particularly when there are other contractors on the site. And the contractor is not reliant on written instructions and he is not required to mitigate his costs unless so instructed.

Having regard to the restrictions on claims in MF/1 arising from the exclusive remedy provisions of clause 44(4) it is difficult to over-state the importance of the contractor's claims entitlement under clause 25(1) and 25(2).

Clause 25(2) – additional cost caused by suspension

Clause 25(2) states simply that any additional cost incurred by the contractor in complying with either:

- the provisions of clause 25.1, or
- engineer's instructions under clause 25.1.

shall be added to the Contract Price. However, both situations are subject to the test of entitlement imposed by clause 25.4.

Under clause 41.2 profit is allowed on claims made under clause 25.2. The procedures for claiming are set out in clause 41.1.

Clause 25.3 – payment for plant affected by suspension

The purpose of clause 25.3 is to ensure that the contractor does not lose his entitlement to payment when a suspension interferes with normal progress. The clause provides that for any suspension of more than 30 days the contract value of the affected plant shall be included in an interim certificate. The contract value as defined in clause 1.1.i is in effect an assessment of the value of the work done at any particular date. However, to safeguard the interests of the purchaser in respect of plant held off-site as a result of suspended delivery the clause also provides that the contractor is not entitled to a certificate of payment until:

- he has marked the plant as the purchaser's property under clause 37.2 (marking of plant);
- he has insured the plant in accordance with clause 47.1 (insurance of works) as if it were on site.

Clause 25.4 – disallowance of additional cost or payment

Clause 25.4 cuts down the apparently open entitlement of the contractor to additional cost under clause 25.2 and interim payment under clause 25.3 by stating that neither is due if the suspension is necessary by reason of:

- default on the part of the contractor, or
- proper execution of the works or plant, or
- safety of the works or plant.

However by the final provisions in the clause the contractor's entitlements are restored if the necessity results from:

- any act or default of the engineer or the purchaser, or
- the occurrence of any of the purchaser's risks – see clause 45.1.

These final provisions appear to be superfluous in respect of the contractor's defaults and presumably they are intended to apply mainly to safety and possibly, in appropriate circumstances, to proper execution.

Clause 25.4 is potentially defective in that it fails to disallow costs and payments relating to suspensions specifically provided for in the contract. But it can be argued that such suspensions do not come within the scope of clauses 25.2 and 25.3 as they do not follow instructions of the engineer. But see the comment in section 10.3 below on clause 19.4 of MF/2.

Clause 25.5 – resumption of work, delivery or erection

Clause 25.5 contains five separate provisions which can be summarized as follows:

- power of the engineer to lift a suspension;
- entitlement of the contractor to require the lifting of certain suspensions lasting more than 90 days;
- entitlement of the contractor to terminate the contract (or omit parts) if certain suspensions last more than 120 days;
- obligation of the contractor to make good any deterioration at the purchaser's cost;
- disentitlement of recovery of costs of making good deterioration in certain circumstances.

Lifting the suspension

The power of the engineer to lift a suspension is expressed in the discretionary term 'may' and not in the mandatory term 'shall'.

No doubt the engineer would see himself as under obligation to lift a suspension as soon as possible. And where the purchaser is responsible for the additional costs arising from the suspension the engineer clearly owes a duty to the purchaser to do so.

However, problems can arise when the suspension results from the contractor's default and the contractor's view of when the default has been remedied allowing the suspension to be lifted differs from the engineer's view. The contractor might see grounds for a claim in such a situation although the contractual basis for such a claim might prove difficult to sustain. See the case of *J. Crosby & Sons Ltd* v. *Portland Urban District Council* (1967) where it was held under an ICE Fourth edition contract that no term could be implied into the contract that the engineer should lift a suspension once the grounds for the suspension ceased to apply and accordingly there could be no claim for damages for breach of such an obligation.

Suspension more than 90 days

Clause 25.5 provides that if a suspension has continued for more than 90 days – and the suspension is not necessitated by the reasons stated in clause 25.4 – the contractor may give notice to the engineer requiring him to give notice to proceed within 30 days.

For the contractor this is a reasonable provision since an indefinite suspension could affect his business generally. For the engineer and/or purchaser 90 days may not be adequate and there is the possibility of the contractor proceeding with the next provision of clause 25.5 which permits termination.

Disappointingly for such an important provision its wording and its application is not as clear as it appears. The difficulty is that the provision applies only to a suspension which is 'not necessitated by the reasons stated in Sub-Clause 25.4'.

The reasons stated in clause 25.4 are default on the part of the contractor or proper execution or safety of the works. The question is does the final proviso of clause 25.4 – covering reasons for which the purchaser is responsible – apply? Consider for example a suspension instructed by reason of safety. If the contractor is at fault then clearly the 90 day provision does not operate. But suppose the safety problem is the responsibility of the purchaser – what then is the position?

The probability is that the intention in such a case is that the 90 day provision does apply – but the wording of the provision may not achieve that result.

Suspension more than 120 days

The third provision of clause 25.5 gives the contractor various options in the event of the engineer failing to give notice to proceed within 30 days of a 90 day suspension after being requested to do so.

The contractor may:

- in respect of affected parts – treat the suspension as an omission variation under clause 27 (variations);
- in respect of the whole of the works – terminate the contract as if under clause 51 (purchaser's default);
- in respect of either parts or the whole elect to do neither of the above – thereby giving entitlement to be paid the contract value of the affected plant.

Making good deterioration

The fourth provision of clause 25.5 requires that the contractor should, on being given notice to proceed, examine the plant and make good any deterioration that has occurred during suspension. The cost of the examination and making good is to be added to the contract price. Profit is allowable under clause 41.2.

Taken by itself the provision would apply to all suspensions, including those necessitated by reason of the contractor's default, but this is qualified by the final provision in clause 25.5 – although not with the best of clarity.

Disentitlement to recovery of costs

The fifth and final provision of clause 25.5 probably intends to say that the contractor shall not be paid the cost of making good to the extent that he is responsible for its deterioration – and that would logically extend to the reasons for the suspension.

What the clause actually says is that the contractor is not entitled to be paid the costs incurred in making good:

- any deterioration, defect or loss caused by defective materials or workmanship, or
- any failure by the contractor to comply with the instructions of the engineer under clause 25.1.

This reference to clause 25.1 presumably relates to the provisions in that clause for storage, preservation and protection. There is, therefore, no stated disentitlement to cost covering the reasons for the suspension.

Clause 25.6 – effect of suspension on defects liability

Clause 25.6 provides for the contractor to recover the additional costs of performing his obligations on defects liability when, solely because of suspension, those obligations are performed more than three years after the normal delivery date.

As with the preceding clause there is no exclusion for suspensions which result from the contractor's defaults.

In practice, however, this may be of no great consequence because only with unusually long suspensions or unusually long periods between delivery and completion will this clause have any effect.

10.3 Delivery and suspension in MF/2

MF/2 includes, with some modification, the provisions of MF/1 on delivery and suspension and also includes additional specialist clauses on delivery.

The relevant clauses are:

- clause 1.1.m   – deferment of delivery
- clauses 18.1 to 18.3 – delivery
- clauses 19.1 to 19.6 – suspension.

Clause 1.1.m – definition of delivery

The definition confirms that delivery terms are to be specified in the special conditions and that delivery, deliver and delivered have the same meaning.

Clause 18.1 – permission to deliver

As with MF/1 the contractor must apply to the engineer for permission to deliver any plant and no plant may be delivered without the engineer's written permission.

Clause 18.2 – delivery terms

This clause confirms that the delivery term applicable to the contract (FOB, CIF, etc) is to be interpreted according to the INCOTERMS rules applying at the date of the contract unless the special conditions state otherwise.

Clause 18.3 – notice to insure

This clause applies only to plant delivered by sea. Its purpose is to retain conformity with INCOTERMS on the obligation of the contractor to give notice to the purchaser to insure and to achieve this it excludes Section 32(3) of the Sale of Goods Act 1979.

Clause 19.1 – Instructions to suspend

Clause 19.1 contains provisions relating to:

- the power of the engineer to suspend;
- deemed instructions on suspension;
- the contractor's obligations to store, preserve and protect during suspension.

These provisions mirror the corresponding provisions in clause 25.1 of MF/1.

Omitted from MF/2 are the MF/1 provisions relating to suspension of erection of plant and the contractor's obligation to maintain his resources on or near the site.

Clauses 19.2 to 19.6

The above clauses match clauses 25.2 to 25.6 of MF/1. The only significant change is that in clause 19.4 of MF/2 cost and payment is disallowed if it relates to a suspension specifically provided for in the programme. Perhaps this would be better worded 'provided for in the Contract'.

10.4 Delivery and suspension in MF/3

The following clauses of MF/3 cover delivery and suspension:

- clause 6.1 – delivery
- clause 8.1 – storage
- clause 9.1 – damage or loss in transit.

Clause 6.1 – place of delivery

The place of delivery is the vendor's works unless another place is named in the contract. For delivery at the vendor's works the vendor shall, if required, load the purchaser's vehicle. For delivery to another place, the vendor shall deliver and the purchaser shall off-load. Property in the goods passes on delivery to the purchaser.

Clause 8.1 – storage

Clause 8.1 covers suspension of delivery and the consequences thereof. The major provisions in the clause are:

- when delivery of goods is delayed for 14 days by the purchaser and:
  - the vendor gives written notice to the purchaser that the goods are ready for delivery, and
  - the goods are suitably and sufficiently marked as appropriated to the contract, and
  - the purchaser is given an opportunity to inspect the goods then the goods are deemed to be delivered in accordance with the contract;
- at the expiration of the 14 day period the goods are at the risk of the purchaser and the amount due on completion becomes due and payable;

- property in the goods does not pass to the purchaser on deemed delivery but passes only on payment;
- the vendor is to store, protect and preserve the goods and insure as required;
- the purchaser is to repay the vendor the cost of storing, protecting, preserving and insuring the goods.

Clauses 9.1 to 9.3 – damage or loss in transit

See section 15.6 of Chapter 15 on accidents, damage and insurances for comment on these clauses.

10.5 *Delivery and suspension in G/90*

In G/90, clause 15 (delivery) and clause 29 (suspension of works) contain the provisions on delivery and suspension. G/90, unlike MF/1, separates delayed delivery and suspension of works but the provisions are to much the same effect.

Clause 15(i) – delivery to site

This clause provides that:

- no plant or contractor's equipment is to be delivered to site until the engineer's authorization has been sought and given;
- the contractor is responsible for delivery, reception and off-loading;
- the contractor is responsible for movement of the plant as required by the specification or directed by the engineer;
- the contractor is entitled to be paid a reasonable sum for movement directed by the engineer.

Clause 15(ii) – delayed plant

For the purposes of clause 15 delayed plant is defined as:

- plant which the contractor is prevented from delivering to the site at:
  - a time specified for delivery, or
  - a time reasonable for it to be delivered (if no time is specified), or
- plant which has been delivered to the site but which the contractor is prevented from erecting, because of
- some delay or failure on the part of the engineer, or some other cause

for which the purchaser or some other contractor employed by him is responsible.

Clause 15(iii) – notice of delayed plant

Clause 15(iii) provides that the contractor may give notice to the purchaser requiring the provisions of clause 15(iv) on delayed plant to have effect when:

- delayed plant is ready for delivery, and
- it has been marked as appropriated to the contract, and
- the engineer has been given the opportunity of inspecting it, or
- it has been delivered to the site.

Clause 15(iv) – consequences of delayed plant

The provisions of clause 15(iv) are stated to operate where notice has been given in accordance with clause 15(iii). Having regard to the decision in the *Crosby* v. *Portland* case mentioned earlier in section 10.2 of this chapter such notice is essential to put the clause into effect.

The provisions are in summary:

- clause 15(iv)(a) – a sum to be added to the contract price for storing, protecting, preserving and insuring delayed plant;
- clause 15(iv)(b) – one month after the normal delivery date the contractor is entitled to have the contract value of delayed plant included in an interim certificate;
- clause 15(iv)(c) – six months after the normal delivery date the contractor is entitled to have 95% of the contract value certified and paid;
- clause 15(iv)(d) – 12 months after the normal delivery date the contractor is entitled to give notice requiring the purchaser to assume responsibility for the plant and is entitled to payment of 100% of the contract value;
- clause 15(iv)(e) – the obligations of the contractor for defects liability do not apply to any defect occurring three years from the normal delivery date or the date of notice given under clause 15(iii) whichever is the later.

Clause 15(v) – responsibility for delayed plant

Clause 15(v) provides that if responsibility for delayed plant passes to the purchaser it does not revert to the contractor until 30 days after notice to proceed or the contractor resumes possession whichever is the earlier.

Clause 15(vi) – defects in delayed plant

Clause 15(vi) states in effect that the contractor is to make good any deterioration in delayed plant.

Clause 15(vii) – costs arising from delayed plant

Clause 15(vii) details the contractor's entitlements to be paid the costs arising in consequence of delayed plant. These are:

- the costs of examining and making good
- the extra expenses of delivering and erecting
- the extra expenses of delayed testing
- the extra expenses of remedying defects.

The first of these is expressly subject to the exception that the contractor's entitlement reduces in so far that the deterioration is due to faulty workmanship or materials or the contractor's failure to protect and preserve.

Clause 29(i) – suspension of works

As discussed in section 17.7 of Chapter 17 on claims for additional payment clause 29(i) of G/90 is one of the most important clauses in the contract.

It provides that if there is a suspension of the works or if the contractor is prevented or delayed from proceeding with the works by the purchaser or the engineer or some other contractor employed by the purchaser then the contractor is entitled to be paid the additional expense. Providing always that the contractor has given notice within a reasonable time of his intention to make a claim.

Clause 29(ii) – prolonged suspension

Clause 29(ii) deals with prolonged suspensions and the consequences thereof. The clause provides that if a suspension exceeds three months the contractor is entitled to have the contract value included in an interim certificate.

Note that unlike MF/1 there is no provision in G/90 for the contractor to terminate the contract either in respect of delayed delivery or prolonged suspension.

Chapter 11

Programmes, delays and damages

11.1 Introduction

A contractor's performance is measured by the purchaser and by the engineer, not just in terms of the technical capability of the plant provided but also by the manner in which the works are executed. Contractors who perform to programme and on time attract repeat business; contractors who fail in these respects frequently disrupt the purchaser's plans; become embroiled in claims; and are likely to be regarded as unreliable.

The manner in which a contract deals with the contractor's obligations to carry out and complete the works and his liability for defaults is therefore of keen interest to all parties. The contract, by its provisions, sets the scene for the judgment of the contractor's performance.

Programmes

The modern trend in contracts is to place increasing emphasis on programmes, method statements and other organizational matters. In some recent contracts such as the New Engineering Contract this is developed into a policy which gives project management as prominent a role in the contract as the expression of the obligations and liabilities of the parties.

The model forms for plant contracts do not go as far as that but all except MF/3 have some programme requirements. And in the 1988 change from MF/A to MF/1 evidence of the trend towards greater programme control can clearly be seen.

Commencement

In a strict legal sense a contract commences when the contractor's offer to undertake the works is accepted by the purchaser. But in most plant, process and construction contracts there is a distinction between the date of commencement of the contract and the date for commencement of the works. The latter date starts time running for the purpose of completion obligations.

Contracts adopt various methods of fixing the date for commencement of the works but not all do so with the precision the matter deserves.

The model forms for plant contracts are reasonably clear and precise with MF/1 and MF/2 covering the point in the definitions section and MF/3 and G/90 relying on the date of acceptance of tender (or placing of order).

Progress

It is debatable whether having been set a date of commencement and a date for completion the contractor's rate of progress in between the two should be of any concern to the purchaser or the engineer. The point was argued in the case of *Greater London Council v. Cleveland Bridge & Engineering Co* (1986) where it was held that as a general principle, in the absence of specific provisions on progress, it is for the contractor to plan and perform his work as desired during the contract period. The contractor's primary obligation is to complete the work within the contract period. If he does that he cannot be said to have failed to exercise due diligence and expedition.

More recently in the case of *Pigott Foundations Ltd* v. *Shepherd Construction Ltd* (1994), it was held that a piling sub-contractor was entitled to plan and perform his work as he pleased provided that he finished within the time allowed in the sub-contract.

Most contracts do, however, contain some obligations on progress, usually stated as obligations to proceed with due diligence and/or expedition. And most state that failure in respect of these obligations is a default which entitles the purchaser to terminate the contract. But, in the light of the above and numerous other legal cases, purchasers should proceed with the greatest of caution before proceeding with termination.

All the model forms for plant contracts except MF/3 impose some obligations in respect of progress but the clauses do not appear to be worded with a great degree of conviction.

Completion

It would seem to go without saying that the procedure for fixing the date for completion in a contract should be clearly stated; that the meaning of completion should be clearly defined; and that provision should be made for the actual date of completion to be clearly marked – ideally by the issue of a certificate to that effect. Frequently, however, these matters are poorly addressed.

In the plant contracts, the taking-over certificate usually marks completion (delivery in the case of MF/3). But, as discussed later in this chapter, there is still plenty of scope for argument on what is meant by completion and how definitive the passing of tests on completion is in confirming that completion has been achieved.

Extensions of time for completion

Extension of time clauses in construction contracts have been well analysed in the courts with the effect that such contracts now generally contain extension provisions drafted with a careful watch on their legal interpretation.

G/90 seems to have followed this approach but MF/1 and MF/2 rather less so. These latter forms contain a potentially serious omission in respect of extending time for sections – a matter which is likely to be rectified in the 1995 Revision 3 of MF/1. However they do contain interesting and somewhat unusual provisions for overcoming or mitigating delay which in principle, if not in the drafting, have a good deal to recommend them.

Liquidated damages for late completion

Liquidated damages can be seen either as compensation for the purchaser for the contractor's default or limitation of the contractor's liability for his default. In the model forms for plant contracts both matters are given consideration.

MF/1, MF/2 and MF/3 all have unusually drafted provisions, which, as with extensions of time, do not apply to sections – although, again, the 1995 Revision 3 of MF/1 will probably rectify this.

G/90 has more conventionally drafted provisions which do apply to sections.

11.2 Programmes generally

Although programmes have long been a common feature of plant, process and construction contracts there are still a great many misconceptions about the role they serve in defining the contractor's obligations and assisting his entitlements to extra payment. This is because programmes are often incorrectly assumed to be contract documents.

Status of programmes

Most contracts call for programmes to be submitted within a specified number of days after acceptance of the contractor's tender. Such programmes will rarely be contract documents. Firstly because it is most unlikely that they will be listed as contract documents; and secondly because their production follows rather than precedes the making of the contract.

Generally only programmes which are submitted with a contractor's tender and are then listed as contract documents, or are otherwise expressly stated to be contract documents, will have the status of contract documents.

Tender programmes and method statements

Until recently few plant, process or construction contracts made any mention of tender programmes or method statements in their standard text. That, however, did not stop, and still does not stop, many purchasers (and/or their engineers) requiring tenderers to include details of their methods and their intended programme with their tenders.

The reasons for this include:

- obtaining information to be used in the appraisal of tenders;
- obtaining information to assist the purchaser and/or engineer in fulfilment of their own obligations;
- obtaining information to assist the purchaser and/or engineer in co-ordinating the contract works with other activities;
- obtaining information in the hope of controlling the contractor's operations and his scope for claims.

The value of this information to the purchaser and/or engineer does depend upon whether it is formally incorporated into the contract so as to bind the appointed contractor. But the danger of incorporating tender programmes and method statements into contracts is that they bind both parties not just the contractor.

Thus in the case of *Yorkshire Water Authority* v. *Sir Alfred McAlpine & Sons (Northern) Ltd* (1985) it was said by Mr Justice Skinner (of an ICE fifth contract into which a tender method statement had been incorporated):

'In my judgment, the standard conditions recognise a clear distinction between obligations specified in the contract in detail, which both parties can take into account in agreeing a price, and those which are general and which do not have to be specified pre-contractually. In this case the [Employer] could have left the programme and methods as the sole responsibility of the [Contractor] under clause 14(1) and clause 14(3). The risks inherent in such a programme or method would then have been the [Contractor's] throughout. Instead, they decided they wanted more control over the methods and programme than clause 14 provided. Hence clause 107 of the specification; hence the method statement; hence the incorporation of the method statement into the contract imposing the obligation on the [Contractor] to follow it save in so far as it was legally or physically impossible. It therefore became a specified method of construction by agreement between the parties.'

Contract programmes

Programmes which are submitted by the contractor after the award of the contract are not, as explained above, contract documents.

The effect of this is that they do not normally create contractual obligations for either party – unless the contract itself expressly provides otherwise. For this, one looks to see whether or not the contract refers in its text to the programme. In the ICE Conditions there is such a reference in relation to possession of the site. And in the IChemE model forms the references to the programme are widely spread and the effect is that the purchaser under those forms is bound to do everything necessary on his part to meet the requirements of the contractor's programme. Additionally the contractor is required to use his best endeavours to perform to his programme and to submit a revised programme if progress is not maintained.

In MF/1 and MF/2 the purchaser acquires programme-related obligations under clause 11 in respect of wayleaves, consents, licences and foundations. And the contractor may be ordered to revise his programme if progress does not match the programme. MF/3 does not mention programmes and G/90 merely requires the contractor to provide a programme and, if so required by the engineer, to vary it.

Approval of programmes

It is often thought that if the contractor's programme is approved by the engineer the approval itself in some way gives the programme contractual status and makes it binding on the parties. Contractors are more inclined to advance this proposition than purchasers and frequently do so to support claims for delay and disruption.

However, most contracts expressly make the point that approvals or consents by the engineer do not relieve the contractor of any of his obligations under the contract – for example, clause 14.3 of MF/1. And it is not obvious why there should be an implied term that approval binds the purchaser. Perhaps the most that can be said is that although approval of a programme will not ordinarily create additional contractual obligations, it may be a factor in determining what is a reasonable time for the performance of obligations.

Shortened programmes

Programmes showing completion in a shorter time than that allowed in the contract frequently cause alarm to engineers who suspect such programmes are more likely to be used as a platform for claims than as a guide to intended progress.

The matter was considered in the case of *Glenlion Construction Ltd* v. *The Guinness Trust* (1987) where the judge was asked to rule whether there was an implied term (in a JCT 63 contract) that where the programme showed early completion, then the employer, through his architect, was obliged to perform so as to enable the contractor to carry out the work in accordance with the programme and to complete early.

The judge held that such a term could not be implied and he made the point that one party cannot unilaterally change the obligations of the other party.

However, it must be emphasized that this decision related only to an implied term and the case does not apply to contracts where there are express terms relating obligations to programmes such as are found in MF/1, MF/2 and the IChemE model forms.

In short, the *Glenlion* case rules out claims on shortened programmes based on implied terms but leaves open the question of the impact of shortened programmes on contractual provisions.

Programmes as the basis of assessment for delays and costs

Because programmes only create contractual obligations when they are either incorporated as contract documents or are expressly linked with obligations in the contract they do not ordinarily form the legal basis of claims for delay and extra cost.

Surprising as it may seem this can work to the benefit of contractors as well as to purchasers. In the case of *Walter Lawrence & Sons Ltd* v. *Commercial Union Properties (UK) Ltd* (1984) a dispute arose as to whether the contractor was entitled to an extension of time for inclement weather.

The architect in correspondence had said: '... It is our view that we can only take into account weather conditions prevailing when the works were programmed to be put in hand, not when the works were actually carried out ...'

The contractor refuted this and claimed that his progress relative to programme was not relevant to his entitlement to an extension. It was held by the judge that the effect of the exceptionally inclement weather was to be assessed at the time when the works are actually carried out and not when they were programmed to be carried out.

However, programmes are useful as evidence in establishing proof of delay and disruption; but even here they are not always the best evidence.

That is found by comparing progress during the delay/disruption period with actual progress during a non-affected period. A method judicially approved in the case of *Whittal Builders Ltd* v. *Chester-le-Street District Council* (1985).

Contractor's failure to produce or revise a programme

If a contract requires the contractor to produce and submit for approval a programme within a specified time and requires the contractor to revise his programme when instructed to do so it will, on the face of it, be a breach of contract if the contractor fails to do either.

The problem then is – does the purchaser have any legal remedy for the contractor's breach or does the engineer have any powers under the contract to oblige the contractor to comply?

As a general rule there are few legal remedies or contractual powers. The purchaser's legal remedies for breach depend on proof of loss and this would be difficult to establish for such a breach. Specific contractual remedies and powers addressing the problem are conspicuously absent from most contracts. In extreme cases it might be possible to terminate the contractor's employment under the contract for default – for example, under clause 49.1(c) of MF/1 on the grounds of neglecting to carry out his obligations under the contract – but the circumstances would have to be appropriate to the gravity of such an action.

An exception to the general rule is found in the New Engineering Contract. Under the terms of that contract, the contractor is entitled to receive only 50% of the amounts due on interim payments until an approved programme is in place.

11.3 Programmes in MF/1

One of the significant changes between the old MF/A and MF/1 was the amount of detail added to the provisions on programmes. The situation now is that MF/1 and MF/2 have comprehensive requirements on programmes but G/90, following MF/A, has only minimal requirements. MF/3 has none at all.

Programmes in MF/1

The term 'programme' is defined in clause 1.1.r of MF/2. 'Programme means the programme referred to in Clause 14 (Programme).'

The benefit of having this defined term for the programme, which appears to say very little, is that it avoids uncertainty as to which programme applies in other clauses of the contract which refer to the programme; most importantly to those parts of clause 11 which link certain of the purchaser's obligations to the timescale of the programme.

Clause 14.1 – Programme

Within the time stated in the Contract or, if no time is stated, within 30 days after the Letter of Acceptance, the Contractor shall submit to the Engineer for his approval the Programme for the execution of the Works showing:

   (a) the sequence and timing of the activities by which the Contractor proposes to carry out the Works (including design, manufacture, delivery to site, erection and testing),

   (b) the anticipated numbers of skilled and unskilled labour and supervisory staff required for the various activities when the Contractor is working on Site,

   (c) the respective times for submission by the Contractor of drawings and operating and maintenance instructions for the approval thereof by the Engineer,

   (d) the times by which the Contractor requires the Purchaser
      (i)   to furnish any drawings or information,
      (ii) to provide access to Site,
      (iii) to have completed any necessary civil engineering or building work (including foundations for the Plant),
      (iv) to have obtained any import licences, consents, wayleaves and approvals necessary for the purposes of the Works,
      (v) to provide electricity, water, gas and air on the Site or any equipment, materials or services which are to be provided by the Purchaser.

The phrase 'the Contractor shall submit' makes it mandatory for the contractor to submit a programme. It is not dependent upon a request from the engineer.

The information to be shown in the programme falls into three categories:

- operational proposals
- submissions to the engineer
- timing obligations on the purchaser.

There is no express requirement to include the activities of subcontractors but the phrase 'the execution of the Works' makes it clear that all activities of any substance must be included.

Clause 14.2 – form of programme

The Programme shall be in such form as may be specified in the Special Conditions or, if not so specified, as may reasonably be required by the Engineer.

Bar charts and histograms remain the most popular and understandable forms of presentation for planning and resource details. For complex projects more sophisticated forms may be appropriate.

Clause 14.3 – approval of programme

Approval by the Engineer of the Programme shall not relieve the Contractor of any of his obligations under the Contract.

This clause may appear to be superfluous in that it is not within the power of the engineer to amend the terms of the contract. However it is useful in establishing the point that approval by the engineer is for the purposes of management and supervision of the contractor's activities and not for the purpose of redefining the contractor's obligations.

As to non-approval or rejection of the contractor's programme by the engineer, MF/1 says nothing. This is a potentially serious omission and clause 14 would be improved by the inclusion of provisions of the type introduced into the ICE Conditions sixth edition for resubmission after rejection of any programme.

In the event of the engineer neither approving nor rejecting the programme it is probably deemed to be approved within a reasonable time after submission. The contract does not expressly say this but neither does it specify any time for the engineer's response and this suggests that the engineer has only a reasonable time in which to approve or reject.

Clause 14.4 – alteration to programme

The Contractor shall not without the Engineer's consent make any material alteration to the approved Programme.

This clause must be read in conjunction with the provision in clause 1.4 that any consent required of the engineer shall not be unreasonably withheld.

It certainly cannot be the intention of the clause that the engineer should be able to prevent the contractor making alterations to the programme which the contractor deems necessary in order to fulfil his obligations or for practical or commercial reasons.

The true intention of the clause is probably that the contractor should keep the engineer fully informed of all proposed changes to the programme.

Clause 14.5 – revision of programme

If the Engineer decides that progress does not match the Programme, he may order the Contractor to revise the programme. The Contractor shall thereafter revise the Programme to show the modifications necessary to ensure completion of the Works within the Time for Completion.

If modifications are required for reasons for which the Contractor is not responsible, the Cost of producing the revised Programme shall be added to the Contract Price.

This clause does not actually require the contractor to work to his programme. But the engineer can order the programme to be revised if there is a departure.

It is suggested that the engineer should not call for a revised programme until he has dealt with any applications for extension of time under clause 33. This is to ensure that the contractor cannot claim that he has been ordered to accelerate by being required to modify the programme to complete earlier than necessary.

It is an odd feature of clause 14.5 that there is no requirement for the contractor to seek approval of the engineer for any revised programme. This can lead to some confusion as to what is meant by the 'approved Programme' in clause 14.4. The intention is probably that approval should be sought for a revised programme in order for it to become the approved programme and an amendment to that effect is sometimes made.

The provision for the cost of producing a revised programme to be added to the contract price is unusual as an express entitlement – but, then in MF/1, the contractor relies on express entitlements for all additional payments.

The difficult aspect of the provision is to determine precisely what is meant by modifications 'for which the Contractor is not responsible'. Clearly if the purchaser or the engineer have, by their actions, made modifications necessary, the contractor is not responsible. But if the cause is a neutral event for which the contractor is entitled to an extension of time the contractor may also be able to argue that he is not responsible for any modification to programme which follows.

11.4 Programmes in MF/2 amd MF/3

MF/2

The programme provisions of MF/2 are essentially the same as those in MF/1 and they are similarly numbered clauses 14.1 to 14.5.

The only changes of note are that in MF/2 the word 'Contract' replaces 'Works'; 'Time for Delivery' replaces 'Time for Completion'; and the following provisions of MF/1 are omitted:

- clause 14.1(b) – numbers of labour and staff
- clause 14.1(d)(ii) – access to site

- clause 14.1(d)(iii) – foundations
- clause 14.1(d)(v) – site services.

MF/3

There are no programme requirements in MF/3.

11.5 Programmes in G/90

'Programme' is not a defined term in G/90. Clauses 8(ii) and 8(iii) are the full extent of the provisions on programmes.

Clause 8(ii) – submission of programme

> Within the period stated in the Appendix after the date of placing the order for the Works the Contractor shall submit to the Engineer a programme showing the sequence and timing of the manufacture and delivery of the Plant (or if the Works are to be completed in Sections of any Section thereof) and the sequence and timing of the part of the Works to be performed on the Site (or any Section thereof) provided always however that in determining the sequence and timing of the Works (or any Section thereof) the Contractor shall have regard to the requirements of the Contract in respect of those matters.

As with MF/1 and MF/2 the submission of a programme to the engineer is mandatory.

There is nothing in G/90, however, to suggest that the programme is to be submitted for the purposes of approval or that the engineer has any control over the form and content of the programme. The intention seems to be no more than to inform the engineer of the contractor's proposals.

Clause 8(iii) – contractor to vary the programme

> The Engineer may at any time by an instruction in writing require the Contractor to vary the programme referred to in Sub-Clause (ii) of this clause either in respect of the sequence or the timing of the Works (or any Section thereof).

This clause supplements clause 10(1) – variations – which gives the engineer power to vary the Works 'including the sequence and timing thereof'. It is probable therefore that an instruction under clause 8(ii) is always to be taken as a variation order – and valued accordingly.

Consequently the apparently wide power of the engineer to require the contractor 'to vary the programme' needs to be exercised with the greatest care since it amounts to power to vary the works.

11.6 Commencement, progress and completion generally

Although the obligations of a contractor to commence, proceed and complete on time are in essence the same whatever the type of contract there is a surprising amount of variation in the way these obligations are expressed even in the limited field of plant and process contracts. And this is not simply a lack of consistency in terminology – confusing as that is – but it is also a matter of what stage of progress has to be achieved to constitute completion.

For example, the model forms for plant and for process contracts require the contractor to complete on time or pay damages for late completion. But none of these forms defines precisely what is meant by completion; and whereas in MF/1 and MF/2 completion is after the taking-over tests; in the IChemE model forms completion precedes the taking-over tests.

It is therefore necessary when dealing with any contract to look carefully at the particular wording to see how the following matters are covered:

- the commencement date
- the time for completion (or the date for completion)
- commencement obligations
- progress obligations
- the meaning of completion
- the certification of completion.

Commencement

Strictly speaking the contract commences when the contractor's tender is accepted by the purchaser. And this date may be used to fix some of the contractor's obligations – for example, the submission of the programme.

However in many contracts the date for commencement of the works is separately specified or left to be fixed by the engineer. It will then normally be from the date for commencement of the works that the time for completion starts to run.

It is essential, therefore, that if there is a time for completion in a contract (rather than a date for completion) that the contract provides a mechanism for specifying or fixing the date for commencement.

MF/1 and MF/2 deal with this by specifying in clauses 1.1.m and 1.1.n respectively the events which trigger the date from which time runs.

MF/3, in clause 7.1, provides a flexible and potentially uncertain arrangement. G/90, in its appendix, indicates that time runs from the placing of the order.

Commencement obligations

Unlike some other standard forms of contract, the model forms for plant do not contain any express requirements that the contractor shall proceed with the works upon the fixing of the date for commencement. However, there are obligations in respect of progress in these forms and any conspicuous failure by the contractor to commence could be evidence of a default.

Progress obligations

The contractor's primary obligation is to complete on time but a contract may impose other obligations in respect of progress. For example, to proceed with due diligence and expedition; to use best endeavours to mitigate delay.

Not infrequently, default in respect of secondary obligations can have more serious consequences for the contractor than the default of failing to complete on time. For the latter the contractor usually pays only damages for late completion whereas failing to proceed with due diligence may be a specified default permitting the purchaser to terminate the contract. This is the case in MF/1, MF/2 and G/90 but as mentioned in section 11.1 above, the purchaser should proceed with the utmost caution.

The meaning of completion

When, as in the plant and process model forms, completion is not a defined term, the meaning has to be ascertained from the contract as a whole or the identification of an event which signifies that completion has been achieved.

In MF/1 and G/90 it is the issue of a taking-over certificate which signifies completion. And as to what has to be achieved before the issue of a taking-over certificate both forms have similar requirements:

- the works shall have passed the tests on or before completion, and
- the works shall be complete except in minor respects which do not affect their use.

Certification of completion

Completion itself, as indicated above, is not certified in the model forms for plant.

The certificate which serves the purpose of the completion certificate is the taking-over certificate and it is that certificate which is effective in establishing when the contractor has fulfilled his obligations in respect of completion.

11.7 *Commencement, progress and completion in MF/1*

Clause 1.1.m – the date for commencement

'Time for Completion' means the period of time for completion of the Works or any Section thereof as stated in the Contract or as extended under Sub-Clause 33.1 (Extension of Time for Completion) calculated from whichever is the later of:

(a) the date specified in the Contract as the date for commencement of the Works;

(b) the date of receipt of such payment in advance of the commencement of the Works as may be specified in the Contract;

(c) the date any necessary legal, financial or administrative requirements specified in the Contract as conditions precedent to commencement have been fulfilled.

It is an unusual feature of MF/1 that the standard appendix to the General Conditions does not list either the time for completion nor the date for commencement. These need to be covered in the Special Conditions or elsewhere in the contract documents.

Care needs to be taken in the drafting of letters of acceptance to ensure that provision (c) of clause 1.1.m does not accidentally come into play. Phrases such as 'subject to contract' and the like may be construed as conditions precedent on general legal principles even if they are not expressly stated to be conditions precedent.

Clause 32.1 – time for completion

Subject to any requirement under the Contract for the completion of any Section before the whole of the Works, the Contractor shall so execute the Works that they shall be complete and pass the Tests on Completion (but not the Performance Tests, if any be included) within the Time for Completion.

This clause supports clause 1.1.v (tests on completion) in making it clear that the tests on completion are to be made before taking-over.

If sectional completion is required to be a contractual obligation it must be specified as such. To be effective a section should be given:

- clear identity
- its own time for completion
- its own damages for late completion.

It is understood that the 1995 Revision 3 to MF/1 will contain additional special conditions for use where the contract is to provide sectional completion and damages for delay in the completion of sections.

Clause 14.6 – rate of progress

> The Engineer shall notify the Contractor if the Engineer decides that the rate of progress of the Works or of any Section is too slow to meet the Time for Completion and that this is not due to a circumstance for which the Contractor is entitled to an extension of time under Sub-Clause 33.1 (Extension of Time for Completion).
>
> Following receipt of such a notice the Contractor shall take such steps as may be necessary and as the Engineer may approve to remedy or mitigate the likely delay, including revision of the Programme. The Contractor shall not be entitled to any additional payment for taking such steps.

Provisions of this kind have been in civil engineering contracts for many years. They are often criticized for placing a duty on the engineer which would be better expressed as a power.

The problem is the phrase 'The Engineer shall notify the Contractor'. This leaves the engineer with no discretion as to whether to act or not. The danger is that if the engineer does notify the contractor that progress is too slow to meet the time for completion and the contractor subsequently establishes an entitlement to extension of time which was effective at the time the notice was given, the contractor may be able to claim that he has been ordered to accelerate. This is sometimes known as constructive acceleration. However, under MF/1, because of the exclusive remedy provisions in clause 44.4 the contractor may have difficulty in finding a legal or contractual basis on which to found his claim.

However, notwithstanding the difficulties the contractor may face in making a claim the engineer should never give notice under this clause until all extension of time applications have been carefully reviewed.

The requirement that the contractor should take steps to remedy or mitigate delay at his own cost is of doubtful effectiveness. The question is what if the contractor prefers to proceed at his own pace and pay damages for late completion? Is the contractor denied this option?

The practical answer is that usually there is little the purchaser or the engineer can do if the contractor ignores the notice to remedy or mitigate delay short of invoking the default provisions of the contract. And if it were to come to this it is suggested the purchaser would be better advised to rely on a default which matched exactly the wording of clause 49 (contractor's default) than rely on a failure under clause 14.6. However, in MF/1 and MF/2 the provisions for prolonged delay discussed later in this chapter might well serve as encouragement to the contractor to comply with the engineer's notice to expedite progress.

Clause 29.2 – taking-over certificates

When the Works have passed the Tests on Completion and are complete (except in minor respects that do not affect their use for the purpose for which they are intended) the Engineer shall issue a certificate to the Contractor and to the Purchaser (herein called a 'Taking-Over Certificate'). The Engineer shall in the Certificate certify the date upon which the Works passed the Tests on Completion and were so complete.

The Purchaser shall be deemed to have taken over the Works on the date so certified. Except as permitted by Clause 30 (Use before Taking-Over) the Purchaser shall not use the Works before they are taken over.

The taking-over certificate serves the purpose of the completion certificate in some other contracts. However it requires both completion and the passing of the tests on completion.

There is no requirement that the contractor should apply for a taking-over certificate when he considers one to be due. The obligation to assess the situation rests with the engineer. This clause is discussed further in Chapter 12.

11.8 Commencement, progress and delivery in MF/2 and MF/3

MF/2 and MF/3 both refer to 'Delivery' where MF/1 refers to 'Completion'.

Commencement in MF/2

Clause 1.1.n applies. This is much the same as clause 1.1.m of MF/1 except that 'the date of the Letter of Acceptance' takes the place of 'the date specified in the Contract'.

Progress in MF/2

Clause 14.6 applies. This is the same as clause 14.6 of MF/1.

Delivery in MF/2

This is a defined term in clause 1.1.m:

> 'Delivery' means delivery in accordance with the delivery terms specified in the General Conditions and 'deliver' and 'delivered' shall have a corresponding meaning.

There is no equivalent of a taking-over certificate in MF/2 and the contractor's obligation to deliver on time as expressed in clause 22.1 is a factual matter.

Clause 22.1:

> The Contractor shall so execute the Contract that the Plant shall be delivered within the Time for Delivery.

Provisions in MF/3

Clause 7.1 applies:

> Any time fixed by the Contract for delivery shall run from the acceptance of the Vendor's tender (or, if there be no tender, of the Purchaser's order) or from the date on which the Vendor is placed by the Purchaser in possession of such information and drawings as may be necessary to enable him to put the work in hand, whichever may be the later. Any time described as an estimate shall not be construed as a time fixed by the Contract.

It might be better if time was simply stated to run from acceptance of the tender.

The provision that time can run from the purchaser's order in the absence of a tender raises the interesting question of how an order can, of itself, make a contract. Where then is the offer which has been accepted?

As to the provision for time starting to run from the provision of necessary information this is inviting dispute on what is necessary.

It is not clear what the reference to 'time described as an estimate' is meant to convey unless it is simply saying that time described in a tender as an estimate is of no contractual significance.

11.9 Commencement, progress and completion in G/90

Commencement

Table B of the appendix to G/90 makes it clear that time for completion runs from the date of placing of the order for the works.

Progress

The only provision in G/90 which can be construed as an obligation on progress additional to that to complete on time is in clause 12 (contractor's default). This permits the purchaser to give the contractor notice to remedy any failure to proceed with due diligence and expedition and power to terminate the contract for continuing default.

Time for completion

The time for completion is to be stated in the appendix in weeks. Clauses 8(iv) and 8(v) state the contractor's obligations in respect of time.

Clause 8(iv):

> The delivery of the Plant (or any Section thereof) shall be effected by the Contractor by the expiration of the relevant Delivery Period shown in the Appendix or by any extended date granted by the Engineer under and subject to Clause 25 (Extension of Time for Completion).

Clause 8(v):

> The Works (or any Section thereof) shall be completed by the expiration of the relevant Time for Completion shown in the Appendix or by any extended date granted by the Engineer under and subject to Clause 25 (Extension of Time for Completion).

Completion

Clause 27(vi) deals with tests:

> Tests before Completion are deemed to form part of the Works and notwithstanding any other requirement for completion, the Works (or any section or portion of the Works) shall not be regarded as complete whilst such tests remain to be passed.

Taking-over certificate

The requirement for the engineer to issue a taking-over certificate on completion is given in clause 28(i):

> As soon as the Works have been completed in accordance with the Contract (except in minor respects that do not affect their use for the purpose for which they are intended and save for the obligations of the Contractor under Clause 30 (Defects after Taking Over) and have passed the Tests before Completion the Engineer shall issue a certificate (herein called a 'taking-over certificate') in which he shall certify the date on which Works have been so completed and have passed the said tests and the Purchaser shall be deemed to have taken over the Works on the date so certified, but the issue of a taking-over certificate shall not operate as an admission that the Works have been completed in every respect.

11.10 Extensions of time generally

The contractor is under a strict duty to complete on time except to the extent that he is prevented from doing so by the purchaser or is given relief by the express provisions of the contract. The effect of extending time is to maintain the contractor's obligation to complete within a defined time and failure by the contractor to do so leaves him liable to damages, either liquidated or general, according to the terms of the contract.

In the absence of extension provisions, time is put at large by prevention and the contractor's obligation is to complete within a reasonable time. See *Dodd* v. *Churton* (1897) and *Peak Construction* v. *McKinney Foundations* (1970).

When time is put at large the contractor's liability is only for general damages; but first it must be proved that the contractor has failed to complete within a reasonable time. See *Rapid Building Group Ltd* v. *Ealing Family Housing Association* (1984).

Purposes of extension provisions

Extension of time clauses, therefore, have various purposes:

- to retain a defined time for completion;
- to preserve the purchaser's right to liquidated damages against acts of prevention;
- to give the contractor relief from his strict duty to complete on time in respect of delays caused by specified neutral events.

Relevant events

The events which entitle the contractor to extensions of time are commonly called 'relevant events'.

These vary in scope and description from contract to contract according to the policy of the particular contract on risk sharing in respect of delays. For example, some contracts allow extensions of time for strikes whereas others do not. However all contracts should include for delays caused by the purchaser or his engineer to ensure that time can be extended in respect of such delays.

'Neutral events' is a term generally used to describe events over which neither party has direct control such as weather and strikes. Where neutral events are included in a contract as relevant events it is sometimes said that the loss lies where it falls. That is, the contractor carries the costs of delay in completing the works whilst the purchaser carries the costs (or losses) he suffers from the late completion. For legal comment on this see the case of *Henry Boot Construction Ltd* v. *Central Lancashire Development Corporation* (1980).

11.11 Extensions of time in MF/1 and MF/2

MF/1 and MF/2 use identical provisions for extension of time except that 'delivery' replaces 'completion' in MF/2. An unusual feature of both model forms is that the provisions apply only to the 'Works' and are not drafted so as to include for 'Sections'.

It could perhaps be argued that by making provision for taking-over by sections in clause 29.1 it is implied that the extension of time provisions extend to sections. But against this it has to be recognized that the consistent approach of the courts over the last two centuries on the interpretation of extension of time and liquidated damages provisions has been to exclude implied terms.

However, note again the comments made earlier in this chapter that the 1995 Revision 3 to MF/1 is likely to expressly deal with sections by additional special clauses.

Clause 33.1 (MF/1) – extension of time for completion

If, by reason of any variation ordered pursuant to Clause 27 (Variations) or of any act or omission on the part of the Purchaser or the Engineer or of any industrial dispute or by reason of circumstances beyond the reasonable control of the Contractor arising after the acceptance of the Tender, the Contractor shall have been delayed in the completion of the Works, whether such delay occurs before or after the Time for Completion, then provided that the Contractor shall as soon as reasonably practicable have given to the Purchaser or the Engineer notice of his claim for an extension of time with full supporting details, the Engineer shall on receipt of such notice grant the Contractor from time to time in writing either prospectively or retrospectively such extension of the Time for Completion as may be reasonable.

Relevant events are grouped into four categories as delays caused by:

- variations
- acts or omissions of the purchaser or the engineer
- industrial disputes
- circumstances beyond the reasonable control of the contractor.

Act or omission

The phrase 'act or omission on the part of the Purchaser' may not be wide enough to cover all delays for which the purchaser could be liable – for example, delays caused by the purchaser's other contractors. However, since any such delay could be covered by the category of circumstances 'beyond the control of the contractor' the omission is probably not serious.

Industrial disputes

'Any industrial dispute' is wide enough to cover disputes both within and without the contractor's organization. To the extent that the contractor's own management practices may have contributed to an industrial dispute it is questionable whether the wording is too generous to the contractor.

Circumstances beyond the contractor's control

'Circumstances beyond the reasonable control of the Contractor' are limited in clause 33.1 to those arising after acceptance of the tender. This is not as straightforward a limitation as it might appear. Its interpretation turns on what is meant by circumstances.

For example, on one interpretation the contractor would have no entitlement to an extension of time for unexpected site conditions which existed prior to acceptance of the tender if the conditions themselves rather than the discovery of the conditions are to be regarded as the 'circumstances'.

Another aspect to consider of the phrase 'circumstances beyond the reasonable control of the Contractor' is how wide is its scope. A similar phrase was given an unexpectedly wide meaning by the House of Lords in *Scott Lithgow Ltd* v. *Secretary of State for Defence* (1989) on another matter. In that case it was ruled that a provision that the contractor should be paid for the effect of 'exceptional dislocation and delay arising during the construction of the vessel due to alterations, suspensions of work or any other cause beyond the contractor's control', extended to failure by suppliers or sub-contractors in breach of the contractual obligations to the contractor.

In clause 33.1 the phrase used is 'beyond the reasonable control' and this may, if anything, widen rather than restrict its scope.

As to delays caused by force majeure as defined in clause 46.1 of MF/1 that clause clearly contemplates that extensions of time will be awarded and presumably, since clause 33.1 does not expressly mention *force majeure*, such delays come under circumstances beyond the reasonable control of the contractor.

Adverse weather

There is no specific relevant event in clause 33.1 for delays caused by adverse weather but it would certainly be open to the contractor to argue that these were beyond his reasonable control.

Delays after the time for completion

Clause 33.1 is more explicit than the extension of time provisions in most other contracts in expressly stating that it applies to delays which occur after the time for completion. By that is meant, after the time the contractor should have completed and the contractor is proceeding in what is sometimes known as culpable delay and is liable for damages for late completion.

The legal position in respect of delays after the time for completion was examined, albeit in the context of the standard building form of contract, JCT 1980, in the case of *Balfour Beatty Building Ltd* v. *Chestermount Properties Ltd* (1993). It was held that the extension of time provisions did apply to such delays and that the extensions granted should be on a net basis rather than on a gross basis.

This latter point disposed of the proposition which had been advanced for many years in respect of many contracts that if the purchaser caused a delay after the time for completion had elapsed then any extension granted should be not just for the period of actual delay but also the intervening period from the completion date up to the delay.

The point is considered to be of general application and to be applicable to plant and process contracts as well as to construction contracts.

For MF/1 and MF/2 (and also G/90) the provisions for delays after the time for completion are so worded that they extend to neutral delays as well as to delays caused by the purchaser. The *Balfour Beatty* case left open the question of whether the contractor had an entitlement to an extension of time for delays caused by such neutral events as strikes and floods which would have been avoided if the contractor had completed on time.

Notice of claim

Clause 33.1 contains a proviso on the duty of the engineer to grant such extensions of time as are due. It reads: 'then providing that the Contractor shall as soon as reasonably practicable have given . . . . . . notice of his claim for an extension of time . . . . . . the Engineer shall on receipt of such notice . . . grant . . . such extension . . . as may be reasonable'.

This may be intended to act as a condition precedent to the granting of any extension of time and no doubt it can be argued that if the contractor fails to give notice he loses his entitlement. But there is a measure of illogicality in this to the extent that the extension of time provisions are intended to operate to protect the purchaser's position. It certainly cannot be the intention of the clause that the contractor has the option of putting time at large by deciding whether or not to give notice.

It may therefore be that the proviso applies only to the granting of extensions 'from time to time' without necessarily ruling out a final review by the engineer (or the purchaser) of the contractor's entitlement. Some judicial support for this approach can be found in the case of *London Borough of Merton* v. *Stanley Hugh Leach Ltd* (1985) although the wording of the contract in that case was admittedly different from that in MF/1.

'The Purchaser or the Engineer'

Further support for limiting the scope of the proviso on notice may be found in the fact that the notice of claim may be given to 'the Purchaser or the Engineer'. This seems to indicate that the purchaser has a potential role in the granting of extensions of time.

'Prospectively or retrospectively'

These words are used to overcome difficulties which have arisen in some contracts on the retrospective granting of extensions of time.

Clearly it is desirable that whenever possible the contractor should be granted extensions of time prospectively so that he has a completion date to aim for and can plan accordingly. It can even be argued that the words 'on receipt of such notice the Engineer shall' imposes a duty on the engineer to deal with claims for any extension with reasonable promptitude – if perhaps, not 'forthwith' as required by some contracts.

'As may be reasonable'

This phrase gives the engineer some flexibility in determining the amount of an extension of time in circumstances such as concurrent delay where the contractor's entitlement may not match exactly the recorded delay.

Clause 33.2 – delay by sub-contractors

> Any delay on the part of a Sub-Contractor which prevents the Contractor from completing the Works within the Time for Completion shall entitle the Contractor to an extension thereof provided such delay is due to a cause for which the Contractor himself would have been entitled to an extension of time under Sub-Clause 33.1 (Extension of Time for Completion).

This is not a general provision entitling the contractor to claim an extension of time for any delay caused by a sub-contractor. Such a provision would cut right across the basic principle that the contractor is responsible for the performance of his sub-contractors.

Clause 33.2 merely provides that the contractor retains his entitlement to extensions of time even though the relevant events apply directly to sub-contractors. In most contracts this is to be implied.

Clause 33.3 – mitigation of consequences of delay

> In all cases where the Contractor has given notice under Sub-Clause 33.1 (Extension of Time for Completion) the Contractor shall consult with the Engineer in order to determine the steps (if any) which can be taken to overcome or minimize the actual or anticipated delay. The Contractor shall thereafter comply with all reasonable instructions which the Engineer shall give in order to overcome or minimize such delay. If compliance with any such instruction shall cause the Contractor to incur extra Costs and the Contractor is entitled to an extension of time under Sub-Clause 33.1, the amount of such extra Costs shall be added to the Contract Price.

There is much good sense in this provision to mitigate delay but the wording could be clearer. The question is – does the contractor get payment instead of an extension of time or does he get both? The point is important in determining whether there is any transfer of risk to the contractor in respect of attempts to mitigate delay. If there is not then the purchaser risks the expenditure of the extra costs without any guarantee of a corresponding benefit.

On strict interpretation of the wording of clause 33.3 it is suggested that the contractor is entitled to receive both an extension of time and payment of extra costs. However there must be some doubt as to whether this is intended. The procedural arrangements of the clause place the initiative with the contractor and this seems odd if he is simply spending the purchaser's money.

The contractor to consult – in all cases

By the wording of clause 33.3 the contractor is obliged to consult with the engineer whenever an application is made for an extension of time. This applies both to delays for which the purchaser is responsible and to neutral delays. Although the contractor does not have discretion on whether or not to consult he must in practice have discretion on what he proposes.

Contractor to comply with all reasonable instructions

In so far that the engineer gives instructions which match or are developed from the contractor's proposals they would no doubt be deemed to be reasonable instructions. For instructions not agreed in the consultation process these might well be deemed to be unreasonable.

Note that the contractor's obligation is only to comply with reasonable instructions – not all instructions. But in any event the obligation may be more theoretical than real because if the contractor refuses to comply with instructions, whether reasonably given or unreasonably given, there is no obvious contractual remedy short of invoking the termination procedures for his default.

Contractor's entitlement to extension of time

The key question for the contractor in deciding whether or not to comply with the engineer's instructions is likely to be – will payment be made for the extra costs? For the answer to this the contractor needs to know whether or not the engineer accepts that there is an entitlement to extension of time.

Unfortunately for the contractor the wording of clause 33.3 does not require the engineer to state his position in advance of giving instructions – although since it is the purchaser's money which is at stake, the engineer would be well advised to know his position and to have obtained the approval of the purchaser before giving any instructions.

Acceleration measures

As a general point it is questionable whether instructions on acceleration type matters should ever be left to the engineer. Usually any payments to the contractor for attempting to finish before the due date are agreed directly between the purchaser and the contractor and both parties know where they stand before acceleration measures are instructed and commenced.

By this test, clause 33.3 of MF/1 is far from satisfactory. To complicate matters further, clause 33.3 also requires the contractor to comply with instructions to overcome or minimize delay for which there is no entitlement to payment - a matter which is already covered in the contract in clause 14.6.

Procedural arrangements

To alleviate the more obvious difficulties of clause 33.3 it is suggested that the following procedure should be adopted:

- contractor applies for an extension of time giving appropriate details;
- contractor indicates to the engineer what steps could be taken to overcome or minimize the delay;
- engineer forms a view on whether any extension of time is due;
- if no extension of time considered due the engineer informs the contractor accordingly and asks for the contractor's proposals under clause 14.6;
- if an extension considered due the engineer consults with the purchaser on the financial implications of taking acceleration measures;
- if the purchaser consents to additional expenditure the engineer instructs the contractor under clause 33.3.

11.12 Extensions of time in MF/3

MF/3 has simple extension of time provisions in clause 7.2 which detail only the relevant events and say nothing on procedural arrangements.

Clause 7.2 – extension of time

> If by reason of instructions or lack of instructions from the Purchaser or by reason of variations ordered under Clause 13 (Variations) or by reason of any industrial dispute or any cause beyond the reasonable control of the Vendor delivery of the goods shall be delayed, the above mentioned time for delivery shall be extended by such period as may be reasonable.

The relevant events are essentially the same as those listed in clause 33.1 of MF/1 and no further comment is given here.

In the absence of any procedural arrangements in the contract for applications and granting of extensions of time they become a matter of agreement between the parties. In the event of disagreement the matter could be referred to arbitration.

11.13 Extensions of time in G/90

Clause 25 of G/90 contains the same provisions as in clauses 33.1 and 33.2 of MF/1 with some slight but not significant differences in wording.

G/90 does not provide for mitigation of delay but it does provide for Sections.

Clause 25 – extension of time for completion

If, by reason of any industrial dispute or any cause beyond the reasonable control of the Contractor (including any impediment, obstruction, delay or default for which the Purchaser is responsible and any variation to the Works (save and except any variation made necessary by the default of the Contractor) directed by the Engineer under Clause 10 (Variations and Omissions) arising after the acceptance of the tender, the Contractor shall have been delayed or impeded in the completion of the Works (or any Section thereof) to the Site, whether such delay or impediment occur before or after the relevant Time for Completion stated in the Appendix or any extended time therefor granted by the Engineer, provided that the Contractor shall without delay have given to the Engineer notice in writing of his claim for an extension of time, the Engineer shall on receipt of such notice as soon as is reasonable grant to the Contractor from time to time in writing either prospectively or retrospectively such extension of the time fixed by the Contract for the completion of the Works (or any Section thereof) or for the delivery of the Plant (or any Section thereof) as may be appropriate and reasonable. Any delay on the part of a Sub-Contractor which prevents the Contractor from completing the Works (or any Section thereof) or delivering the Plant (or any Section thereof) by the relevant Time for Completion or the expiry of the relevant Delivery Period as the case may be shall entitle the Contractor to an extension of the appropriate time provided such delay was due to any cause for which the Contractor himself would have been entitled to an extension of time under this clause.

Little needs to be added here to the comments on clauses 33.1 and 33.2 of MF/1 except to draw attention to the requirement in clause 25 that the engineer should grant extensions of time 'as soon as is reasonable'. This is not stated in MF/1 although it can probably be implied.

11.14 Damages for late completion generally

The purchaser's entitlement to damages for the contractor's default of late completion may be specified in the contract as either liquidated or unliquidated.

Unliquidated damages

Damages which are assessed after the breach according to general legal principles are called unliquidated damages. Proof of the loss suffered is essential to the recovery of such damages.

Liquidated damages

The essence of liquidated damages is a genuine covenanted pre-estimate of loss. The characteristic of liquidated damages is that loss need not be proved.

Most standard forms of contract are drafted to permit the parties to fix in advance the damages payable for late completion. When these damages are a genuine pre-estimate of the loss likely to be suffered, or a lesser sum, they can rightly be termed liquidated damages. In short, liquidated damages are fixed in advance of the breach whereas general, or unliquidated damages, are assessed after the breach.

Reasons for use

There are sound commercial reasons for using liquidated damages. They bring certainty to the consequences of breach and they avoid the expense and dispute involved in proving loss.

Liquidated damages provisions are not solely for the benefit of the purchaser. They are also beneficial to contractors for they not only limit the contractor's liability for late completion to the sums stipulated, but they also indicate to the contractor at the time of his tender the extent of his risk. Thus, if a contractor believes that he cannot complete within the time allowed he can always build into his tender price his estimated liability for liquidated damages. All that the purchaser gets out of liquidated damages is relief from the burden of proving his loss and usually the right to deduct liquidated damages from sums due to the contractor.

Exhaustive remedy

It needs to be emphasized that if liquidated damages provisions are valid, they provide an exhaustive and exclusive remedy for the specified breach.

This was one of the points confirmed in the case of *Temloc* v. *Errill Properties Ltd* (1987) where it was said:

> 'I think it clear, both as a matter of construction and as one of common sense, that if ... the parties complete the relevant parts of the appendix ... then that constitutes an exhaustive agreement as to the damages which are ... payable by the contractor in the event of his failure to complete the works on time.'

The effect of this is that the purchaser cannot choose to ignore the liquidated damages provisions and sue for general damages. Nor can he recover any other damages for late completion beyond those specified.

Limitation of liability

One important aspect of liquidated damages as applied to plant and process contracts is that they operate as a limitation of the contractor's liability.

In the case of *Widnes Foundry Ltd* v. *Cellulose Acetate Silk Co Ltd* (1933) the contractor was not prepared to take the financial risk of any greater sum than £20 per week for late completion and this was inserted in the contract. When a dispute arose on final payment after a 30 week delay in delivery and erection the purchaser counterclaimed not the £600 due as liquidated damages but £5850 as unliquidated damages. It was held that the contractor's liability was limited to £600.

In the *Temloc* case mentioned above the limitation effect was even more apparent. In that case, liquidated damages had been entered into the contract as £NIL. And it was held that £NIL was the effective sum for the calculation of the liquidated damages due.

If the provisions fail

It is well settled that when a liquidated damages clause fails to operate because it is successfully challenged as a penalty, or fails because of some defect in legal construction, act of prevention or other obstacle, then general damages can be sought as a substitute. Thus it was said in *Peak* v. *McKinney* (1970):

> 'If the employer is in any way responsible for the failure to achieve the completion date, he can recover no liquidated damages at all and is left to prove such general damages as he may have suffered.'

There is no firm ruling in English law that liquidated damages invariably act as a limit on any general damages which may be awarded in their place although this is generally thought to be the position.

Liquidated damages and penalties

There is a good deal of misunderstanding on the relationship between liquidated damages and penalties. It is often thought that if the purchaser cannot prove his loss then any sum taken as liquidated damages must be a penalty. This is certainly not the case.

Providing the sum stipulated as liquidated damages is a genuine pre-estimate of loss it is immaterial whether or not the loss can be proved or even suffered; a point well illustrated by the case of *BFI Ltd* v. *DCB Integration Systems Ltd* (1987).

However, if the stipulated sum is extravagant relative to any likely loss it may be held by the courts to be a penalty. English law does not allow the recovery of penalties and the courts will look at the stipulated sum irrespective of whether it is called liquidated damages or a penalty and if it is found to be a penalty will limit damages to the proven amount flowing from the breach. It is this which encourages contractors and their lawyers to find ways of challenging liquidated damages as penalties.

The rules on distinguishing penalties from liquidated damages laid down by Lord Dunedin in *Dunlop Pneumatic Tyre Co Ltd* v. *New Garage and Motor Co Ltd* (1915) are founded on the principle that liquidated damages must be a genuine pre-estimate of loss. Although an extravagant sum is the most obvious target for challenge many cases have succeeded before the courts on technical arguments on whether or not the stipulated sum could, in all circumstances, be a genuine pre-estimate of loss.

Such arguments will have less chance of success in the future. The Privy Council held in the case of *Phillips Hong Kong Ltd* v. *The Attorney General of Hong Kong* (1993) that:

- The time when the clause should be judged was at the time of making of the contract not the time of the breach.
- So long as the sum payable was not extravagant having regard to the range of losses which could be reasonably anticipated, it can be a genuine pre-estimate of the loss.
- The argument that, in unlikely and hypothetical situations, the clause might provide damages greater than the loss actually suffered, was not valid.
- What the parties have agreed should normally be upheld.

The decision in the above case was followed in *J. F. Finnegan Ltd* v. *Community Housing Association Ltd* (1993) where it was held that the use of a formula to calculate the amount of liquidated damages was justified.

11.15 Liquidated damages in MF/1

The provisions for liquidated damages for late completion in MF/1 are unusual in many respects. They appear not to have been drafted with precise regard to the well established legal rules for such damages. And although they may well protect the contractor from liabilities for unliquidated damages they could well, if challenged, prove ineffective in providing the purchaser with a means of compensation for the contractor's default.

However, by far the most unusual aspect of liquidated damages in MF/1 (and MF/2) is that in the case of prolonged delay when the limit of ordinary liquidated damages stated in the contract has been exhausted additional unliquidated damages start to become payable.

Clause 34.1 – delay in completion

> If the Contractor fails to complete the Works in accordance with the Contract, save as regards his obligations under Clauses 35 (Performance Tests) and 36 (Defects Liability), within the Time for Completion, or if no time be fixed, within a reasonable time, there shall be deducted from the Contract Price or paid to the Purchaser by the Contractor the percentage stated in the Appendix of the Contract Value of such parts of the Works as cannot in consequence of the said failure be put to the use intended for each week between the Time for Completion and the actual date of completion. The amount so deducted or paid shall not exceed the maximum percentage stated in the Appendix of the Contract Value of such parts of the Works, and such deduction or payment shall subject to Sub-Clause 34.2 (Prolonged Delay), be in full satisfaction of the Contractor's liability for the said failure.

'Fails to complete the Works'

As with the extension of time provisions in MF/1 there is nothing here to cover delay in the completion of Sections. This is in marked contrast to the provisions of the IChemE model forms which apply not only to sections but to any other specified thing.

But again the point is made that the 1995 Revision 3 to MF/1 will probably deal with liquidated damages for sections.

'If no time be fixed'

The arrangement in clause 34.1 that liquidated damages become due if the contractor fails to complete within a reasonable time 'if no time be fixed' is most unusual. Normal convention is that liquidated damages apply only to fixed times and unliquidated damages apply when time is at large.

But even if legal obstacles are not fatal to the application of liquidated damages to unspecified times there is the practical point to resolve of who decides whether or not the contractor has failed to complete within a reasonable time.

This alone should be sufficient to make the point that entering into a contract with no fixed time for completion will rarely be in the purchaser's interests.

The percentage stated in the appendix

The appendix of MF/1, in common with other plant and process model forms, requires the rate of liquidated damages to be stated as a percentage of the contract value for each week of delay.

The rate suggested in the official 'Commentary' on MF/1 published by the sponsors of the model form is between $\frac{1}{4}$% and 1% per week.

If the rate selected is equivalent to, or less than, the purchaser's genuine pre-estimate of loss then it can genuinely stand as liquidated damages. But otherwise the rate may be challengeable as a penalty.

In practice the application of a financing formula of the type approved in the *Finnegan* case mentioned in section 11.13 gives a rate in the order of $\frac{1}{2}$% per week when modest supervision charges are included. And for the majority of contracts the purchaser's true and full pre-estimate of loss would probably greatly exceed 1% per week.

Damages per week

It should be noted that if liquidated damages are expressed per week they cannot be apportioned down to give a rate per day. Consequently, damages per week only apply to whole weeks of delay.

'Such part of the Works'

These words suggest the intention that the rate of liquidated damages should be applied not so much to the contract price of the whole of the works as to the contract value of parts of the Works.

Two construction cases, *Bruno Zornow (Builders) Ltd* v. *Beechcroft Developments Ltd* (1990) and *Turner* v. *Mathind* (1986) illustrate the difficulty of applying liquidated damages to parts of the works which are not specified as sections with their own rates of damages. Those cases did admittedly deal with specified sums rather than percentages as liquidated damages but they show that the courts will not enforce liquidated damages provisions which are not fully specified.

The problem with MF/1 in referring to parts of the works is who is to specify the value of the parts in question and how is the difficulty that there is no provision for extension of time for parts to be overcome.

Cannot be put to the use intended

The words 'cannot in consequence of the said failure be put to the use intended' introduce yet more uncertainty into clause 34.1.

The said 'failure' is the failure of the contractor to complete the works. That in itself is all that is needed to activate the contractor's liability for late completion. Either the contractor has got a taking-over certificate or he has not. It is therefore difficult to see why failure to complete is qualified by consideration of whether or not the works, or parts of the works, cannot in consequence be put to their intended use.

This seems to suggest that the contractor can challenge his liability for liquidated damages on the grounds that although he failed to complete in time the purchaser was not ready for them. Again, this cuts across the basic legal principle that with liquidated damages the purchaser is not required to prove his loss.

Time for completion

This term is defined in clause 1.1.m and it does include for time as extended under clause 33.1.

The amount not to exceed the maximum percentage

The appendix to MF/1 requires a maximum percentage to be stated for liquidated damages – thus putting a gross limit on the contractor's liability. The official Commentary on MF/1 suggests a maximum of between 5% and 15%. This is purely a commercial arrangement and there is no legal requirement for any maximum limit. So the question of what maximum percentage should be used has to be decided on the commercial circumstances of the parties in each contract. The purchaser may get the benefit of better competitive prices if the maximum percentage is kept low. However, that needs to be balanced against the potential problems of the purchaser if the contractor is under no financial pressure to complete on time.

'In full satisfaction of the Contractor's liability'

This phrase merely confirms the legal position that liquidated damages are an exhaustive and exclusive remedy. But see clause 34.2 below.

Clause 34.2 – prolonged delay

> If any part of the Works in respect of which the Purchaser has become entitled to the maximum amount provided under Sub-Clause 34.1 (Delay in Completion) remains uncompleted the Purchaser may by notice to the Contractor require him to complete. Such notice shall fix a final Time for Completion which shall be reasonable having regard to such delay as has already occurred and to the extent of the work required for completion. If for any reason, other than one for which the Purchaser or some other contractor employed by him is responsible, the Contractor fails to complete within such time, the Purchaser may by further notice to the Contract elect either:
>
> (a) to require the Contractor to complete, or
> (b) to terminate the Contract in respect of such part of the Works, and recover from the Contractor any loss suffered by the Purchaser by reason of the said failure up to an amount not exceeding the sum stated in the Appendix or, if no sum be stated that part of the Contract Price that is properly apportionable to such part of the Works as cannot by reason of the Contractor's failure be put to the use intended.

The purpose of clause 34.2 is to keep the contractor under pressure to complete when the maximum amount of liquidated damages for late completion has been reached. The purchaser is, in effect, given power to make time of the essence (for which the remedy for defaults is termination of the contract) by serving notice on the contractor.

'Any part of the Works'

These words re-introduce the complications in clause 34.1 on the application of the provisions to parts of the Works. But the concept of termination of the contract in respect of parts is even more legally dubious than that of applying liquidated damages to parts.

'A final Time for Completion'

It is apparently left to the Purchaser (not the engineer) to fix the final time for completion. The words used suggest that this time cannot later be extended.

Subsequent delays

It is far from clear how clause 34.2 operates if the purchaser or some other contractor employed by him is responsible for delay after the notice fixing a final time for completion is served. The probability is that the provisions for termination lapse and the purchaser has no further remedy.

It would also seem from the specific words 'Purchaser or some other Contractor' that the risk of all other causes of delay rests with the contractor and that no relief is given after the fixing of the final time for completion for the effects of neutral events.

The purchaser may elect

By further notice, after the final time for completion has elapsed, the purchaser may either:

- require the contractor to complete, or
- terminate the contract in respect of parts.

The purchaser is not under a duty to serve this further notice – the clause reads 'the Purchaser may'. The contractual position if the purchaser fails to serve any further notice would seem to be the same as if the purchaser serves notice requiring the contractor to complete. Namely the purchaser is prepared to accept completion whenever it is achieved.

There is nothing in the clause to suggest that the purchaser has the power to fix a second final time for completion thereby re-activating his option to terminate the contract.

Termination of the contract

There is no reference in clause 34.2 linking termination under the clause with the provisions for termination under clause 49 (contractor's default). Nor is there anything in clause 49 referring back to clause 34.2.

This suggests that termination under clause 34.2 is wholly independent of the principal termination provisions of MF/1 as set out in clause 49. The consequences of termination under clause 34.2 are therefore purely those described in the clause.

And these consequences are that the purchaser may:

- terminate the contract in respect of the delayed parts, and
- recover unliquidated damages up to the amount stated in the appendix.

This, of course, is after the contractor has already exhausted the limit of liquidated damages stated in the contract. This combination of liquidated damages and unliquidated damages for late completion is most unusual.

'Any loss suffered by the Purchaser'

This phrase may not mean what it says. Clause 44.2 of the contract prevents recovery of indirect and consequential damage except in relation to clauses 34.1 and 35.8.

An amount not exceeding the sum in the appendix

The appendix requires the maximum loss recoverable by the purchaser to be stated as a sum of money. There is nothing to stop this sum exceeding the contract price but the intention is probably that if a sum is stated it is less than the contract price.

If no sum stated

Where no sum is stated in the appendix the maximum liability of the contractor under clause 34.2 is the contract price. But this applies only when the whole of the works cannot be put into use as intended. Where only part of the works cannot be put into use an apportionment must be made.

In other words, if no maximum sum is stated in the appendix, the contractor's maximum liability for any part of the works which is still not capable of being used after prolonged delay is the proportion of the contract price that the price of the part has to the whole.

Payment for work completed

As a general rule, damages for late completion are deductible from the contract price. Thus the contractor retains his entitlement to payment for work completed notwithstanding that he completes late and is liable for a deduction for damages.

In the case of damages payable for prolonged delay where the termination option is exercised, the contractor may not have entitlement to payment of any of the contract price but he remains liable nevertheless for the purchaser's loss.

11.16 Liquidated damages in MF/2

Clause 24.1 – delay in delivery

If the Contractor fails to deliver the Plant or any Section thereof within the Time for Delivery, or if no time be fixed, within a reasonable time, there shall be deducted from the Contract Price or paid to the Purchaser by the Contractor the percentage, stated in the Appendix, of the Contract Value of such parts of the Plant as cannot in consequence of the said failure be put to the use intended for each week between the Time for Delivery and the actual date of delivery, but the amount so deducted or paid shall not exceed the maximum percentage stated in the Appendix of the Contract Value of such parts of the Plant, and such deduction or payment shall subject to Sub-Clause 24.2 (Prolonged Delay), be in full satisfaction of the Contractor's liability for the said failure.

The comments on clause 34.1 of MF/1 apply generally here.

Clause 24.2 – prolonged delay

If any part of the Plant in respect of which the Purchaser has become entitled to the maximum amount provided under Sub-Clause 24.1 (Delay in Delivery) remains undelivered, the Purchaser may by notice to the Contractor require him to deliver the same within such time (not being less than 28 days) as the Purchaser may specify in the notice. If the Contractor shall fail to deliver the undelivered Plant within the time so specified the Purchaser shall be entitled, after having given the Contractor notice of his intention to do so, to purchase plant in place of the undelivered Plant and there shall be deducted from the Contract Price that part thereof which is properly apportionable to the unde-livered Plant. The Contractor shall pay to the Purchaser any sums by which the expenditure reasonably incurred by the Purchaser in obtaining plant in place of the undelivered Plant exceeds the sum deducted. All plant obtained by the Purchaser in place of undelivered Plant shall comply with the Contract and shall be obtained at reasonable prices and where practicable under competitive conditions.

Prolonged delay under MF/2 appears to follow the same general prin-ciples as MF/1 in that the purchaser can give notice fixing a final delivery time and, if this is not achieved, can then obtain from other sources any parts still outstanding.

However, the wording of clause 24.2 creates a significant difference between the model forms as to the consequences of prolonged delay.

Whereas in MF/1 the purchaser can recover his losses (up to specified maximum amounts) in MF/2 all that the purchaser can recover is the cost of obtaining the non-delivered parts. And even that is subject to the purchaser obtaining the parts at reasonable prices and where practicable under competitive conditions.

As sanction on the contractor for prolonged delay this amounts to next to nothing and the purchaser might be better advised to look to his remedies under clause 34 (contractor's default) than apply the provisions of clause 24.2.

11.17 Liquidated damages in MF/3

Clause 7.3 – delay in delivery

> If the Purchaser shall have suffered any loss by the failure of the Vendor to deliver goods in accordance with the Contract within the time fixed thereby or any extension thereof or, if no time be fixed, within a reasonable time, the Purchaser shall be entitled to recover liquidated damages from the Vendor. Such damages shall be a sum equal to the percentage stated in the Appendix of that part of the Contract Price which is properly apportionable to such portion of the goods as cannot in consequence of such failure be put to the use intended for each week until the Vendor has delivered goods in accordance with the Contract, or goods in replacement have been provided by the Purchaser pursuant to Sub-Clause 5.2 (Consequence of Rejection). Provided always that the amount so recoverable shall not exceed the maximum percentage specified in the Appendix. In default of specification in the Appendix the percentages above mentioned shall be one-half per cent and five per cent respectively.

'If the Purchaser shall have suffered any loss'

These are odd words to see in a provision for liquidated damages. Such damages normally apply without proof of loss and even when it can be shown by the contractor that the purchaser has suffered no loss. See the case of *BFI Group of Companies Ltd* v. *DCB Integration Systems Ltd* (1987).

Clause 7.4 – prolonged delay

> When the sum recoverable by the Purchaser as liquidated damages has amounted to the maximum above provided, the Purchaser shall be entitled by notice in writing to the Vendor to require him to deliver the goods within such time (not being less than 28 days) as the Purchaser may specify in the notice. If the Vendor shall fail to deliver the goods within the time so specified the Purchaser shall, without prejudice to his rights under Sub-Clause 7.3 (Delay in Delivery), be entitled, after having informed the Vendor in writing of his intention so to do, to obtain goods in place of those which the Vendor has failed to deliver and there shall be deducted from the Contract Price that part thereof which is properly apportionable to the undelivered goods. The Vendor shall pay to the Purchaser any sum by which the expenditure reasonably incurred by the Purchaser in obtaining goods in place of undelivered goods exceeds the sum

deducted. All goods obtained by the Purchaser in place of undelivered goods shall comply with the Contract and shall be obtained at reasonable prices and when practicable under competitive conditions.

This follows the provisions of MF/2 and does nothing to recompense the purchaser for the consequences of prolonged delay.

Clause 7.5 – liability for delay

The Purchaser's remedies under Sub-Clause 7.3 (Delay in Delivery) and 7.4 (Prolonged Delay) shall be in lieu of any other remedy in respect of the Vendor's failure to deliver goods in accordance with the Contract within the time fixed thereby or any extension thereof or, if no time be fixed, within a reasonable time.

This clause operates very much to the protection of the vendor and it has the effect of depriving the purchaser of his common law rights for the vendor's default.

11.18 Liquidated damages in G/90

Clause 26 – delay in completion

If the Contractor fails to complete the Works or (if the Works are to be completed in sections) any section of the Works (except in minor respects that do not affect their use for the purpose for which they are intended and save for the obligations of the Contractor under Clause 30 (Defects After Taking-Over) by the relevant Time for Completion stated in the Appendix or within any extended time fixed under Clause 25 (Extension of Time for Completion) and the Engineer certifies in writing that in his opinion the same ought reasonably to have been completed, then the Contractor shall pay or allow the Purchaser a sum calculated at the relevant rate stated in the said Appendix as Liquidated and Ascertained Damages for the period during which the Works or the relevant section thereof shall so remain or have remained incomplete, but not exceeding the upper limit stated therein, and the Purchaser may deduct such sum from any monies due or to become due to the Contractor from the Purchaser or the Purchaser may recover the same from the Contractor as a debt. Provided that if any portion or portions of the Works or any section of the Works is taken over by the Purchaser the sum due shall be reduced by the proportion which the value of the portion of portions taken over bears to the value of the Works or of the relevant section. For the purposes of this Clause the Works (or sections or portions thereof) shall be regarded as complete if a taking over certificate has been issued or becomes due in relation thereto or if the Works (or sections or portions) have been put to beneficial use by the Purchaser otherwise than under the provisions of Clause 28(iii).

This is a more conventional liquidated damages clause than in MF/1, MF/2 or MF/3. It complies with general legal principles and it is clear in its language.

In short, the contractor becomes liable to pay liquidated damages if he fails to complete and the engineer certifies that he should have completed.

Chapter 12

Taking-over

12.1 Introduction

In plant and process contracts the point in time at which responsibility for the plant passes from the contractor to the purchaser is unambiguously called 'taking-over'. Construction contracts lack such a definitive term and transfer of responsibility takes place on 'completion'.

Taking-over and completion are not different concepts. Both occur when the works are substantially complete and are fit to be put into use. But whereas taking-over usually marks both completion of the work to be done and the passing of tests on completion, the achievement of 'completion' in construction contracts is usually no more than an inspection based event.

Taking-over then is the end of the construction stage of the contract and all that remains for the contractor to do in normal circumstances is to finish outstanding minor items of work, to assist with performance tests and to honour his defects liability obligations. There will, of course, be financial and administrative matters to settle but the contractor who has achieved taking-over has fulfilled for most practical purposes his primary obligations to design, manufacture, deliver, erect, test and complete the works.

Contractual effects

For both parties taking-over is of great contractual importance. Its principal effects are:

- the purchaser is free to put the works into use;
- responsibility for care of the works is transferred from the contractor to the purchaser;
- the contractor's liability for damages for late completion ceases to run;
- the defects liability period commences.

For the contractor there are also secondary benefits:

- payments become due;
- retention monies may be released in whole or part;
- insurance responsibilities are reduced;
- bonds and guarantees may (in some cases) be released.

Defining taking-over

Having regard to the contractual significance of taking-over the conditions which have to be satisfied for it to be achieved should be clearly stated in the contract.

There are two aspects to this. Firstly, the specification should set out with clarity details of the tests to be passed and the work to be undertaken prior to completion. Secondly, the conditions of contract should state in general terms, but again with clarity, what is to be achieved before taking-over. In particular whether taking-over follows passing the specified tests, completion of the works or the achievement of both.

In some model forms, including MF/1, there is a degree of uncertainty on this latter point. And although it is unlikely that contracts intend that passing the tests on completion is alone sufficient to warrant taking-over there is a not uncommon view that this is the case.

Usually the problem arises because of words which seem to link completion to passing the tests on completion – giving the impression that passing the tests is completion. Thus in clause 29.2 of MF/1 taking-over occurs when the works 'have passed the Tests on Completion and are complete'; and the date of taking-over is by that clause 'the date upon which the Works passed the Tests on Completion and were so complete'.

However even if it is arguable from the above words that in MF/1 passing the tests on completion constitutes completion there are other provisions in the model form, (clause 32.1 for example) which make it clear that completion and passing the tests on completion are separate issues.

It is suggested that it would require very clear words in a contract for taking-over to automatically become due on the passing of the tests on completion.

Deemed taking-over

Both MF/1 and G/90 use the phrase 'deemed to have taken-over' in connection with taking-over which has been formalised by the engineer in a certificate.

To some extent this is confusing because 'deemed' is normally used to describe circumstances which have not been formalized. As for example, in G/90 where the works are 'deemed to have been taken-over' when there is premature use by the purchaser.

One explanation for the practice in MF/1 and G/90 of referring to deemed taking-over even where a certificate is issued is that it emphasises the obligation of the engineer to act with fairness in issuing a certificate when it becomes due. In other words, when the contractor has fulfilled his obligations the engineer must issue the taking-over certificate whether or not the purchaser is ready to put the works into use. In that sense the purchaser is deemed to have taken over the works whether he does so or not.

There is another point which is that the date of issue of any taking-over certificate will normally be days or weeks later than the date of taking-over in the certificate. In retrospect it might therefore be said that the works were deemed to have been taken-over at the earlier date.

Effects of deemed taking-over

Informal deemed taking-over does not necessarily have the same effects as formal taking-over. If a contract provides that certain events occur on the 'issue' of a taking-over certificate it is arguable as to whether they also occur on informal deemed taking-over.

The two most likely effects of informal deemed taking-over, whether specified in the contract or not, are the transfer of responsibility for care of the works and the commencement of the defects liability period.

The extent to which the contractor's liability for damages for late completion is affected can only be determined from the precise wording of the delay provisions but in most cases damages for delay reduce in proportion to the value of the works put into use – and the value of the works deemed to be taken-over will usually amount to the same thing.

Use before taking-over

Both MF/1 and G/90 state that the purchaser shall not use the works before taking-over except in circumstances where the contractor has failed to complete within one month of the due date. In practice, this is a restriction which cannot always be followed. For example, in a contract to refurbish a live pumping station the plant will be put into use progressively as the work proceeds.

Where the model forms do not expressly cater for use before taking-over but it cannot be avoided the best solution is to include in the special conditions provisions which permit such use or to use a model form such as the BEAMA Conditions mentioned in section 1.3 of Chapter 1 which are particularly intended for reconstruction and repair.

In the absence of such special conditions the contractor has a strong argument that premature use amounts to deemed taking-over.

Taking-over of parts and sections

Both MF/1 and G/90 allow for the taking-over of pre-defined sections of the works.

G/90 also covers the taking-over of parts (or portions) by agreement of the parties. MF/1 says nothing on this but it can still be achieved by the parties agreeing to the designation of part of the works as a section. This can be formalized by varying the specification under clause 27.

Certification of taking-over

For a matter as important as taking-over it is preferable that its occurrence is recorded in a document which leaves no doubt as to its purpose.

MF/1, most unusually amongst model forms of any type, includes a specimen taking-over certificate. This includes a useful schedule covering matters related to taking-over.

The use of the phrase 'deemed to have been taken-over' in the specimen certificate (and elsewhere) has been discussed above. Perhaps it would be better if the phrase 'commercial use' in the specimen certificate matched the phrase in clause 29.2 'use for the purpose for which they are intended'.

Model forms without taking-over

Model forms for supply and delivery such as MF/2 and MF/3 do not have provisions for taking-over. The contractor (or vendor) fulfils his primary obligations on delivery.

12.2 Taking-over in MF/1

Clause 29 of MF/1 deals with taking-over and clause 30 deals with use before taking-over.

Clause 29.1 – taking-over by sections

> If the Contract provides for the Works to be taken over by Sections the provisions of this Clause shall apply to each such Section as it applies to the Works.

Clause 1.1.a defines 'Sections' as the parts into which the works are divided by the specification.

The effect of clause 29.1 is, therefore, that where the specification details sections, those sections can be the subject of separate taking-over under clause 29. Such sections usually have their own times for completion and damages for late completion although, as discussed in Chapter 11, clauses 33 and 34 dealing with these matters do not expressly cover sections.

As to parts of the works or sections not detailed in the specification there is no provision for taking-over under clause 29 even by agreement as in G/90 – but see clause 30.1 on use before taking-over.

Clause 29.2 – taking-over certificates

When the Works have passed the Tests on Completion and are complete (except in minor respects that do not affect their use for the purpose for which they are intended) the Engineer shall issue a certificate to the Contractor and to the Purchaser (herein called a 'Taking-Over Certificate'). The Engineer shall in the Certificate certify the date upon which the Works passed the Tests on Completion and were so complete.

The Purchaser shall be deemed to have taken over the Works on the date so certified. Except as permitted by Clause 30 (Use before Taking-Over) the Purchaser shall not use the Works before they are taken over.

As mentioned earlier in this chapter there is some ambiguity in this clause which implies that passing the tests on completion constitutes completion. However that is not the intention of the contract and taking-over requires both passing the tests on completion and completion.

It is clear from the phrase 'except in minor respects that do not affect their use' that 'completion' does not have to be absolute and is in effect no more than substantial completion.

The actual form of the taking-over certificate is a matter for the engineer. There is no obligation to use the specimen provided in MF/1 although its schedule is helpful confirmation to the parties of associated matters.

Every taking-over certificate should carry at least two dates:

- the date of issue of the certificate
- the date the works were complete and had passed the tests on completion (the date of taking-over).

These dates serve different purposes in relation to care of the works.

Transfer of responsibility for care of the works passes from the contractor to the purchaser on the date of taking-over – clause 43.1. But the joint insurance of the works taken out by the contractor must continue in force until 14 days after the date of issue of the taking-over certificate – clause 47.1.

For defects liability the obligations of the contractor commence on the date of taking-over – clause 36.1 – and not the date of issue of the certificate. A third date which it is advisable to state in taking-over certificates, is the date for completion of outstanding works – see clause 29.4 below.

For comments on use of the works before they are taken over see under clause 30.1 below.

Clause 29.3 – effect of taking-over certificate

Upon the issue of a Taking-Over Certificate, risk of loss or damage to the Works (other than any parts thereof excluded by the terms of the Taking-Over Certificate) shall pass to the Purchaser and he shall take possession thereof.

This is a poorly worded clause which, taken literally, is in conflict with other provisions of MF/1.

Clearly the date of issue of a taking-over certificate is not the date on which the purchaser takes possession of the works. And, as mentioned above, responsibility for care of the works transfers to the purchaser on the date of taking-over, not the date of issue of the certificate.

However, it is understood that the above matters will be rectified in the 1995 Revision 3 to MF/1 so that clause 29.3 will then commence 'With effect from the date of taking-over as stated in the Taking-Over Certificate'.

Clause 29.4 – outstanding work

The Contractor shall rectify or complete to the reasonable satisfaction of the Engineer within the time stated in the Taking-Over Certificate any outstanding items of work or Plant noted as requiring rectification or as incomplete. In the event the Contractor fails to do so, the Purchaser may arrange for the outstanding work to be done and the Cost thereof shall be certified by the Engineer and deducted from the Contract Price.

The specimen taking-over certificate in MF/1 requires outstanding works to be completed within 28 days 'after the date of this Certificate' – which presumably means its date of issue.

For clause 29.4 to be effective, the engineer needs to append a detailed list of outstanding works to each taking-over certificate. It is not unknown for contractors to argue when there is no such list, or no time for completion of outstanding works is specified, that they have until the end of the defects liability period to complete outstanding works. This is an abuse of the defects liability period which is intended only for defects arising during the liability period but nevertheless purchasers are often reluctant to take action until the period has expired.

Even where, as in clause 29.4 the purchaser has an express right to arrange the completion of outstanding works there is a natural reluctance to do so in case the works are accidentally damaged in the process or in case the contractor is provided with an excuse for defects.

Where the purchaser does arrange for the completion of outstanding work the 'cost' the engineer should certify for deduction from the contract price can, by definition of cost in clause 1.1.y, include for all expenses and overheads.

Clause 30.1 – use before taking-over

> If, by reason of any default on the part of the Contractor, a Taking-Over Certificate has not been issued in respect of the whole of the Works within one month after the Time for Completion, the Purchaser shall be entitled to use any Section or part of the Works in respect of which a Taking-Over Certificate has not been issued, provided the same is reasonably capable of being used. The Contractor shall be afforded the earliest possible opportunity of taking such steps as may be necessary to permit the issue of the Taking-Over Certificate. The provisions of Sub-Clause 43.1 (Care of the Works) shall not apply to any Section or part of the Works while being so used by the Purchaser and Clause 36 (Defects Liability) shall apply thereto as if a Taking-Over Certificate had been issued from the date the Section or part was taken into use.

Clause 29.2 of MF/1, quoted above, states that except as permitted by clause 30 the purchaser shall not use the works before they are taken over.

All that clause 30 does is to permit the purchaser to use the works, or any part or section, when the contractor has failed to complete within one month of the due date for completion.

Neither clause 30, nor any other clause in MF/1, addresses the issue (which is, however, faced in G/90) – if the purchaser does use the works or any part or section before the issue of a taking-over certificate are they deemed to have been taken over from the date they were put into use.

The opening words of clause 30.1 'by reason of any default on the part of the Contractor' raise the question – what is the position if the contractor is not in default but there is no further entitlement to extension of time? Perhaps the words assume that if the contractor is not in default he must have entitlement to an extension of time – but that is a doubtful proposition.

The consequences of the purchaser putting the works into use under clause 30.1 are stated in the clause to be:

- the contractor is to be afforded the earliest opportunity of taking steps to permit the issue of the taking-over certificate;
- responsibility for care of the works put into use transfers from the contractor to the purchaser;
- the defects liability period commences from the date the works are put into use.

There is a further consequence in that it would appear from clause 34.1 (delay in completion) that damages for late completion cease to run from the time the works are put into use.

There is an obvious danger that this combination of circumstances removes from the contractor some of the key incentives to complete. It is essential, therefore, that terms of payment are drafted to retain an incentive.

Clause 31.1 – interference with tests

If by reason of any act or omission of the Purchaser, the Engineer or some other contractor employed by the Purchaser, the Contractor shall be prevented from carrying out the Tests on Completion in accordance with Clause 28 (Tests on Completion) then, unless in the meantime the Works have been proved not to be substantially in accordance with the Contract, the Purchaser shall be deemed to have taken over the Works and the Engineer shall, upon the application of the Contractor, issue a Taking-Over Certificate accordingly.

This clause has been discussed in Chapter 9 on testing.

The point to make here is that this is a case where there is deemed taking-over (in the sense that it may not actually have happened) and there is also an obligation on the engineer to issue a taking-over certificate on the application of the contractor.

Note that there are no provisions for taking-over in MF/2 or MF/3.

12.3 Taking-over in G/90

Clause 28 of G/90 deals with taking-over. G/90 does not have a specimen taking-over certificate as does MF/1.

Clause 28(i) – taking-over

As soon as the Works have been completed in accordance with the Contract (except in minor respects that do not affect their use for the purpose for which they are intended and save for the obligations of the Contractor under Clause 30 (Defects after Taking-Over) and have passed the Tests before Completion the Engineer shall issue a certificate (herein called a 'taking-over certificate') in which he shall certify the date on which Works have been so completed and have passed the said tests and the Purchaser shall be deemed to have taken over the Works on the date so certified, but the issue of a taking-over certificate shall not operate as an admission that the Works have been completed in every respect. Save as provided in Sub-Clause (iii) of this clause the Purchaser shall not use the Works or any section or portion thereof until a taking-over certificate has been issued in respect thereof. If nevertheless the Purchaser does so use the Works or any section or portion thereof the Works or section or portion shall be deemed to have been taken over.

This clause makes it clear that taking-over follows both completion and the passing of tests before completion.

Outstanding works are dealt with in negative fashion – the issue of a taking-over certificate does not operate as an admission that the works have been complete in every respect. Curiously there is no express obligation in G/90 for the contractor to actually complete the outstanding work. The positive approach of MF/1 is to be preferred.

The reference to sub-clause (iii) is to use before taking-over in the event of late completion. See below.

Unlike MF/1, G/90 does recognize that the works may be put into use notwithstanding provisions to the contrary. And G/90 does in such circumstance confirm that the works are deemed to have been taken-over – although it does not state that the engineer is obliged to issue a certificate to that effect.

Engineers should be wary of issuing certificates when they are not obliged to do so in case they prejudice the position of the parties in later disputes.

Clause 28(ii) – sections

If the Works are divided into two or more sections, Sub-Clause (i) hereof shall apply to each section as it applies to the Works. If by agreement between the Purchaser and the Contractor any portion of the Works (other than a section or sections) shall be taken over before the remainder of the Works, the Engineer shall issue a taking-over certificate in respect of that portion.

Sections are not defined as a contractual term in G/90 but they are clearly intended to be pre-defined in the specification and the appendix.

The provision for taking-over of portions by agreement is of benefit to both parties. This is missing from MF/1.

Clause 28(iii) – use before taking-over

If by reason of any default on the part of the Contractor a taking-over certificate has not been issued in respect of every portion of the Works within one month after the date fixed by the Contract for the completion of the Works, or if no date be fixed, within a reasonable time , the Purchaser shall be at liberty to use the Works or any portion thereof in respect of which a taking-over certificate has not been issued if and so long as the Works or the portion so used as aforesaid shall be reasonably capable of being used provided that the Contractor shall be afforded reasonable opportunity of taking such steps as may be necessary to permit the issue of the taking-over certificate. The provisions of Clause 21(i) (Liability for Accidents and Damage) shall not apply to any portion of the Works while being so used by the Purchaser.

This is much the same as clause 30.1 of MF/1. There may appear to be a significant difference in that clause 28(iii) does not state that the defects liability period commences when the works are put into use. But this follows in G/90 from clause 30(i)(b).

Clause 28(iv) – interference with tests

If by reason of any act or omission of the Purchaser, or the Engineer, or some other contractor employed by the Purchaser, the Contractor shall be prevented from carrying out the Tests before Completion as provided in Clause 27(i) (Tests before Completion) then, unless in the meantime the Works shall have been proved not to be substantially in accordance with the Contract, the Purchaser shall be deemed to have taken over the Works, and the Engineer shall issue a taking-over certificate accordingly.

This part of clause 28(iv) is identical to clause 31.1 of MF/1.

Chapter 13

Performance testing

13.1 Introduction

Performance tests, as described in plant and process contracts, are tests designed to establish whether the plant operates as specified under working conditions.

In this sense performance tests can readily be distinguished from other tests on the plant because:

- they are made after taking-over when the plant is in use, and
- they are carried out by the purchaser – with the contractor observing or supervising.

In a narrow technical sense there may be no difference between certain performance tests and tests on completion. For example, in the water industry, tests on effluent quality are frequently essential before plant is taken over and put into use. And although such tests may be repeated on a regular basis after taking-over and will then, in the contractual sense, be performance tests they will in the first instance be classed as tests on completion.

Purpose of performance tests

Performance tests serve two functions:

- they show whether the plant is performing to a minimum acceptable standard; and if not, whether modification, or in the extreme, rejection is necessary, and
- they show whether there is any shortfall in the plant's performance which, although within acceptable limits, entitles the purchaser to some monetary recompense – usually set by way of liquidated damages for low performance.

Suitability for performance testing

Not all plant is suitable for performance testing. In fact until the first edition of MF/1 in 1988 the standard model forms for plant did not cover the subject. Even now, there is nothing in MF/2 and MF/3. And although MF/1 and G/90 have provisions on performance testing they operate on an 'if and when included' basis. So it is for the purchaser to decide on each contract whether or not performance tests should be included.

Clearly much depends upon whether the plant can be tested as a single functioning unit or whether it has been integrated into some larger project where its performance can no longer be individually measured to any meaningful purpose. In this, plant contracts are far more variable than process plant contracts where there is normally an output which is capable of analysis and measurement.

But even when the output aspect of a plant contract is not suitable for performance testing it may be possible, and appropriate, to measure certain input characteristics such as power consumption and to include relevant performance tests for these in the contract.

Consequences of failures

Various possible remedies for the failure of performance tests exist according to the seriousness of the failure and the express terms of the contract:

- modification of the malfunctioning plant;
- payment of liquidated damages for low performances;
- acceptance of the plant subject to a reduction in the contract price;
- rejection of the plant on the grounds of unacceptably low performance.

None of these is wholly without its difficulties particularly in relation to timing and compatibility with other contractual provisions. The problem essentially is that the contractor usually has, before he commences the performance tests, a taking-over certificate which signifies that the contractor has fulfilled his obligation to construct and complete the works and which starts time running towards the finality of all his obligations under the contract.

Consequently it is advisable that provisions for performance tests have their own strict timetables and where these are not found in the model forms they should be included in the special conditions or the relevant schedules.

13.2 Performance testing generally

Scope of tests

The tests for each contract are unique depending on the specification requirements; the guarantees given by the contractor; and the purpose of the plant in the particular works.

But amongst the matters commonly subject to performance tests in plant and process contracts are:

- ability of the plant to achieve quoted efficiency
- power consumption
- cost of operating and maintaining the plant
- product quantity and quality
- consumption of chemicals
- quantity and quality of effluents
- volume of waste products
- pollution and noise control.

Setting the parameters

The essence of useful performance testing is that it concentrates only on those things for which measurable performance criteria can be established. Tests must be selective and the number of parameters used must be limited.

The questions to ask at the outset are – how can this be measured and to what degree of accuracy can it be measured? It is pointless to embark on performance testing which simply ends in files of inconclusive data.

The process of testing

Unless the contract states otherwise it will normally be the purchaser's responsibility to carry out and to provide the instrumentation for performance testing.

Ideally the testing should be done using the plant's own control systems and instrumentation. If the contractor is required to provide equipment, manpower or other resources during the testing process it should be specified so that it can be priced in the tender.

Not uncommonly the contractor's rates for supervising performance tests are given on an hourly or daily basis.

Liquidated damages for low performance

The purchaser is only entitled to stipulate liquidated damages for low performance if he can make a genuine pre-estimate of his loss. If the likely loss is too indeterminate to be quantified the damages should be left as unliquidated. This does, of course, raise the point that unliquidated damages are only recoverable on proof of loss. There is, therefore, little point of performance tests for which loss can neither be pre-estimated nor proved.

Most liquidated damages for low performance are related to the expected lifetime of the plant or some lesser time period. Occasionally upper limits on the total of low performance damages are set but more commonly the acceptance parameters themselves fix the limits.

In some circumstances it can be appropriate to balance liquidated damages for low performance with provisions for bonus payments for higher than expected performance but there is no essential legal link between the two.

13.3 Performance testing in MF/1

MF/1 contains provisions for performance testing but they apply only where such tests are included in the contract.

Unlike the IChemE model forms there is no specific numbered schedule in MF/1 for performance tests and they may appear either in the special conditions or in the specification. As with G/90 (see comment later in this chapter) the tests may be expressly linked to performance guarantees provided by the contractor.

Clause 35.1 – time for performance tests

> Where Performance Tests are included in the Contract they shall be carried out as soon as is reasonably practicable and within a reasonable time after the Works, or the Section of the Works to which such tests relate, have been taken over by the Purchaser.

Ideally in MF/1 performance tests should be completed and their outcome finalized within the 12 months defects liability period. If they continue after this there are contractual difficulties unless the model form is amended.

The problem lies in clause 39.12 (effect of final certificate of payment) which makes the final certificate conclusive evidence that the contractor has performed all his obligations under the contract and that the works are in accordance with the contract. And by clause 39.9 (application for final certificate of payment) the contractor is entitled to apply for the final certificate when he has completed any outstanding remedial work at the end of the defects liability period.

Unlike the IChemE model forms of contract, there is no provision in MF/1 for an acceptance certificate to mark the satisfactory passing of performance tests.

Clause 35.2 – procedures for performance tests

> Performance Tests shall be carried out by the Purchaser or the Engineer on his behalf under the supervision of the Contractor and in accordance with the procedures and under the operating conditions specified in the Contract and in accordance with such other instructions as the Contractor may give in the course of carrying out such tests.

This clause confirms that it is the purchaser (or his engineer) who should carry out performance tests but the contractor should have the right to intervene to ensure that the plant is properly operated.

Clause 35.3 – cessation of performance tests

> The Purchaser, or the Engineer on his behalf, or the Contractor shall be entitled to order the cessation of any Performance Test if damage to the Works or personal injury are likely to result from continuation.

This is a sensible provision which recognizes the joint role of the purchaser, engineer and contractor in safety matters and the interests of the contractor in plant for which he is still responsible.

Clause 35.4 – adjustments and modifications

> If the Works or any Section thereof fails to pass any Performance Test (or repetition thereof) or if any Performance Test is stopped before its completion, such test shall, subject to Sub-Clause 35.5 (Postponement of Adjustments and Modifications), be repeated as soon as practicable thereafter. The Purchaser shall permit the Contractor to make adjustments and modifications to any part of the Works before the repetition of any Performance Test and shall, if required by the Contractor, shut down any part of the Works for such purpose and restart it after the adjustments and modifications have been made. All such adjustments and modifications shall be made by the Contractor with all reasonable speed and at his own expense. The Contractor shall, if so required by the Engineer, submit to the Engineer for his approval details of the adjustments and modifications which he proposes to make.

The wording here is slightly odd in stating that performance tests which fail 'shall' be repeated. It is unlikely that this is intended to be mandatory. Probably the intention is that the contractor can require the tests to be repeated if he is dissatisfied with the results or if he thinks that by adjustment or modification he can effect some improvement in performance.

Note that although adjustments and modifications are stated to be at the contractor's expense there is no mention of compensation for the purchaser when the works have to be shut down whilst such adjustments and modifications are made. Nor is there provision elsewhere in MF/1 for such compensation. The most the purchaser can do is require postponement of the adjustments and modifications under clause 35.5. However it is understood that the 1995 Revision 3 to MF/1 will add a new provision into clause 35.4 to the effect that any additional cost incurred by the purchaser solely by reason of the repetition of any performance test shall be deducted from the contract price.

Clause 35.5 – postponement of adjustments and modifications

If the Works or any Section thereof fails to pass any Performance Test (or repetition thereof) and the Contractor in consequence proposes to make any adjustments or modifications thereto, the Engineer may notify the Contractor that the Purchaser requires the carrying out of such adjustments or modifications to be postponed. In such event the Contractor shall remain liable to carry out the adjustments or modifications and a successful Performance Test within a reasonable time of being notified to do so by the Engineer. If however the Engineer fails to give any such notice within one year of the date of taking over of the Works or Section thereof, the Contractor shall be relieved of any such obligation and the Works or Section thereof shall be deemed to have passed such Performance Test.

This clause confirms the point made above that ideally performance tests should be conducted within the defects liability period. Moreover to preserve as much as possible of that period for any postponements required by the purchaser the tests should commence as soon as possible after taking-over. In the event of a test scheduled to take place towards the end of the 12 months after taking-over the purchaser is left with little scope to ask for any postponement.

Clause 35.6 – time for completion of performance tests

If the Contract provides that the Performance Tests (or repetition thereof) shall be completed within a specified time the Purchaser shall be entitled to use the Works as he thinks fit from the expiry of such time.

It is certainly sensible to fix times for the passing of performance tests in the contract (usually in the special conditions) if only to avoid arguments as to when the consequences of failure come into effect.

Clause 35.6, however, appears to imply that if the contract does specify times for the completion of performance tests then until that time has elapsed the purchaser is not entitled to use the works as he thinks fit.

Perhaps what is intended is that once the specified time has elapsed the contractor's rights to make adjustments and modifications are lost and the contractor cannot require a shut down of the works.

Clause 35.7 – evaluation of results of performance tests

> The results of Performance Tests shall be compiled and evaluated jointly by the Purchaser, or the Engineer on his behalf, and the Contractor in the manner detailed in the Contract. Any necessary adjustments to the results to take account of any previous use of the Works by the Purchaser, the measuring tolerances and any differences between the operating conditions under which the Performance Tests were conducted and those detailed in the Specification or performance test schedule shall be made in accordance with the provisions of the Specification or, if the Specification contains no such provisions, then in such manner as is fair and reasonable.

This clause covers an essential aspect of performance testing, namely, that the contract must set out the manner in which tests results are to be compiled and evaluated. Similarly, tolerances and the range of working conditions of the plant need to be detailed in the contract.

Without such precision in the contract, all too often performance testing ends in inconclusive argument over the results.

Clause 35.8 – consequences of failure to pass performance tests

Clause 35.8 covers the whole range of consequences of failure to pass performance tests – from payment of liquidated damages for low performance to rejection of the works for gross failure. MF/1 is unusual in this, and the better for it. Other model forms are less forthright in spelling out the full consequences of failures.

Clause 35.8(a) – liquidated damages

> Where liquidated damages for failure to achieve any guaranteed performance have been specified in the Special Conditions and the results are within the stipulated acceptance limits the Contractor shall pay or allow to the Purchaser

the liquidated damages so specified. Upon payment or allowance of such liquidated damages by the Contractor the Purchaser shall accept the Works.

Liquidated damages should only be applied to failures which fall within the range which can be tolerated.

Usually the payment of such damages is a one-off sum which then relieves the contractor of any future obligation. This is presumably what is meant by 'the Purchaser shall accept the Works'. There is no other formal provision for acceptance in MF/1.

Clause 35.8(b) – reduction in price

Where such damages have been so specified but the results are outside the stipulated acceptance limits, or where liquidated damages have not been so specified, the Purchaser shall be entitled to accept the Works or the Section subject to such reasonable reduction in the Contract Price as may be determined by arbitration under Clause 52 (Disputes and Arbitration).

This clause covers two situations:

● failures beyond stipulated limits, and
● damages where no liquidated damages are specified.

In both cases the payment is a sum of money left to be decided after the breach of contract and it is therefore, in legal terms, unliquidated damages although it is described in the clause as a reduction in the contract price.

It is assumed in clause 35.8(b) that the works are fit to be put to permanent use notwithstanding the performance failures. This follows from the words 'the Purchaser shall be entitled to accept'.

It might appear from these words that the purchaser has the option of whether or not to accept the works but it is apparently not intended that the purchaser should be free to reject the works unless the failure deprives him substantially of the benefit of the works (see clause 35.8.c which follows).

It may therefore be more correct to say that the purchaser is obliged to accept the works in return for payment of unliquidated damages if the works can reasonably be put to permanent use.

This would clearly be a better interpretation of the contractual position where, for example, there was only a marginal failure in the performance tests but the contract failed to specify liquidated damages.

This interpretation of the clause is reinforced by the provision for the reduction in the contract price to be determined in arbitration if not agreed between the purchaser and the contractor.

Clause 35.8(c) – rejection of the works

> Where such failure of the Works or the Section would deprive the Purchaser of substantially the whole of the benefit thereof the Purchaser shall be entitled to reject the Works or the Section and to proceed in accordance with Clause 49 (Contractor's Default).

In the event of a dispute as to whether the purchaser was deprived substantially of the benefit of the works the matter could be dealt with under clause 52 (disputes and arbitration).

The effect of proceeding to clause 49 (contractor's default) would in such circumstances probably be a nil termination value. That would leave the contractor liable for the cost of renewal of the works (subject to any limitations of liability in the contract) without any entitlement to payment for the rejected works.

13.4 Performance testing in MF/2 and MF/3

Neither MF/2 nor MF/3 have specific provisions for performance testing when the plant is in use. In both model forms the only tests covered are tests before delivery, and these are tests which will normally be carried out on the contractor's premises. See clause 17 of MF/2 and clause 4 of MF/3.

Clearly such tests can, and frequently will have, some element of performance written into the specification. But in the context of this chapter these are not performance tests as such for in the event of failure the plant will not be accepted for delivery. Consequently the question of sub-standard performance in use should not arise.

Liquidated damages for low performance

Although neither MF/2 nor MF/3 cover liquidated damages for low performance it is not unusual for the tests carried out before delivery to show the plant to be acceptable even if not quite up to the specified standards.

In anticipation of this many contracts let under MF/2 and MF/3 do have special provisions specifying liquidated damages for low performance which is within acceptable parameters. Sometimes this is done by requiring the contractor to state in his tender the performance guarantees he is offering and the price reductions he will allow for marginal failures.

Requirements for on-site testing

Where the purchaser's requirements are such that on-site performance testing is essential then special conditions have to be added to MF/2 (or

MF/3) covering, for the particular contract, those matters detailed in clauses 35.1 to 35.8 of MF/1.

13.5 Performance testing in G/90

G/90 like MF/1 provides for performance testing on an 'if and when specified' basis. However the wording of clause 27(vii) falls well short of the detail in MF/1.

Clause 27(vii) – performance tests

> If the Contractor has stated in the Contract the performance to be achieved for any Plant, and if, after passing the relevant Tests before Completion, such Plant fails to meet the stated performance and cannot be further modified within the time fixed by the Contract or any extension of such time granted by the Engineer or if no time be fixed within a reasonable time to achieve such performances then damages shall be payable to the Purchaser. If the Specification contains a formula for calculating the amount of damages such formula represents the Purchaser's genuine pre-estimate of the damages likely to be suffered by him in the event that the Plant fails to meet the performance stated in the Contract, and such damages shall be payable by the Contractor as Liquidated and Ascertained Damages.

The opening words of this clause can be improved by substituting 'Contract' for 'Contractor'. As the clause stands it operates only when the contractor has given performance guarantees.

Carrying out the performance tests

Clause 27(vii) does not state which party should carry out performance tests. But since they follow the passing of the 'Tests before Completion' it can be assumed that they are intended to be undertaken after taking-over when the plant is in use and under the control of the purchaser.

Time for performance tests

As with MF/1, performance tests under G/90 should normally be carried out as soon as possible after taking-over and ideally should be completed within the defects liability period.

Although the final certificate under G/90 (available at the end of the 12 months defects liability period) is only stated to be conclusive evidence as to the value of the works it does appear from clause 31(xii) that the contractor's liability for defects after the final certificate is limited to latent defects.

Damages for failure

The only stated remedy in clause 27(vii) for failure of performance tests is given as 'damages shall be payable to the Purchaser'. This appears to suggest that rejection of the plant is not contemplated and that acceptance will follow payment of compensating damages – either liquidated or unliquidated.

However, in the event of gross failure it may be that the scope of 'damages' could extend to rejection of the plant and all the damages which follow such rejection.

Liquidated damages for low performance

The wording of clause 27(vii) on liquidated damages for low performance merely emphasizes that such damages must be a genuine pre-estimate of loss if they are to stand as liquidated damages.

Chapter 14

Liability for defects

14.1 Introduction

The provisions on defects liability in both plant and process contracts are frequently a matter of surprise to people more used to dealing with construction contracts.

Not only are there financial limits on liability in most plant and process contracts but in virtually all such contracts the contractor's liability for defects expires well before the time which would normally apply to latent defects. Added to this there are usually, in such contracts, exclusion clauses which seek to limit or exclude liability for negligence and in some contracts the limitation periods for bringing legal actions are reduced. The overall effect is that the purchaser's ordinary legal rights and remedies are seriously diminished.

The approach of some purchasers to this is to amend the general conditions in the model forms by:

- raising or deleting financial limits on liability;
- amending or deleting provisions which limit latent defects liability;
- amending or deleting exclusion clauses;
- amending or deleting clauses which reduce limitation periods.

Such an approach is wholly understandable when the purchaser has borrowed long term money to finance a long life project and his investment runs to major sums. To do business on a commercial footing the contractor must then be prepared to shoulder liabilities appropriate to his involvement.

However, for many contracts the limitations and exclusions in the model forms are not unreasonable and they may well provide a risk sharing arrangement which is to the commercial benefit of both parties in so far that the contract price reflects risk.

For plant contracts there is the practical point that most defects – some estimate this at 80% to 85% – are revealed within the standard 12 months defects liability period. Moreover, having regard to the responsibilities of the purchaser for operating and maintaining plant after taking-over the question of latent defects is by no means straightforward.

14.2 Limitations and exclusions generally

Firstly it is necessary to distinguish between limitation clauses and limitation periods.

Limitation clauses

These can be either financial limitation or time limitation clauses:

- Financial limitation clauses seek to reduce the damages payable by one party for his breach or default to a specific or ascertainable sum. Thus the guilty party does not deny liability but he seeks to put a ceiling on his liability. In plant contracts this ceiling is usually set at the amount of the contract price for any one default.
- Time limitation clauses do not seek to affect the damages payable but instead they place an agreed limit on the period in which the guilty party's liability can be enforced. In plant contracts this is usually set at 12 months after taking-over.

Both types of limitation clauses are in the general nature of exclusion clauses and as such potentially come under the statutory control of the Unfair Contract Terms Act 1977. However, clauses which seek to limit liability rather than to deny outright any liability are far more likely to satisfy the statutory test of reasonableness in the Act and they will almost certainly be construed more leniently than full exclusion clauses.

Limitation periods

Limitation periods are the periods prescribed by statute for commencing legal proceedings. Under English law the periods are governed by the Limitation Act 1980 and for breaches of contract they are:

- six years for simple contracts
- twelve years for speciality contracts – that is those under seal or executed as a deed.

Under Scottish Law a slightly different approach applies but there are comparable periods of 5 and 20 years.

These limitation periods start to run from the time when the cause of action occurs and that is either:

- the date when the breach of contract occurs, or
- in the case of defective work, the last date when the contractor completed all the work he undertook in the contract (which is usually taken to be the date of taking-over)

Excluding liability for negligence

Exclusion and limitation of liability clauses can purport to exclude or limit liability for both breach of contract and negligence. But special principles of construction apply to clauses which seek to exclude outright liability for negligence.

The law on this is very complex but there is an underlying principle that it is improbable that one party to a contract should have agreed to the loss of his expected contractual benefits when such loss is caused by the other party's negligence or lack of care.

However, in the case of *Ailsa Craig Fishing Co Ltd* v. *Malvern Fishing Co and Securicor (Scotland) Ltd* (1982) the House of Lords held that the principles governing the construction of exclusion clauses as defences to liability for negligence did not apply to clauses which merely limit damages for negligence.

In that case Securicor were under contract to patrol certain vessels moored in Aberdeen harbour. Through admitted negligence the full patrol service was not provided and a vessel which tilted, sank another vessel. Securicor in their defence relied on a limitation clause in the contract which limited damages to £100 per incident or £10,000 in total. The House of Lords held that the clause was effective.

The Unfair Contract Terms Act 1977

Statutory control over exclusion clauses is provided by the Unfair Contract Terms Act 1977. By the Act certain types of clauses are deprived of all effect whilst others are given effect only in so far that they are shown to be reasonable.

The Act commences:

'An Act to impose further limits on the extent to which under the law of England and Wales and Northern Ireland civil liability for breach of contract, or for negligence or for other breach of duty, can be avoided by means of contract terms and otherwise and under the law of Scotland civil liability can be avoided by means of contract terms.'

In broad terms, the Act governs contract terms which:

- exclude or restrict liability for negligence;
- exclude or restrict liability for breach of contract;
- permit different contractual performance from that expected or permit no performance at all;
- require indemnities against the other party's negligence or breach of contract;

- exclude liability for breach of terms implied into contracts by the Sale of Goods Act and the Supply of Goods and Services Act;
- exclude liability in respect of misrepresentation.

Negligence liability under the Act applies only to liability arising from business activities.

The Act governs liability arising in contract when one party either:

- deals as a consumer, or
- deals on the other's written standard terms of business.

Test of reasonableness

In relation to a contract term there is a test of reasonableness in Section 11(1) of the Act to the effect that a term:

> 'shall have been a fair and reasonable one to be included having regard to the circumstances which were, or might reasonably have been, known to or in the contemplation of the parties when the contract was made.'

Reasonableness applied to limitation clauses

If the clause is one of limitation, Section 11(4) of the Act requires that the court must also have regard to the defendant's resources and to the extent to which he might have been able to cover himself by insurance, in determining the reasonableness of the contract term.

The Latent Damage Act 1986

The Latent Damage Act 1986 amends the Limitation Act 1980 to cover situations where latent defects come to light after the normal limitation periods have expired.

The Act was passed to remedy the unacceptable legal position which previously applied and which was highlighted in the case of *Pirelli* v. *Oscar Faber & Partners* (1983). There it was held that the date of accrual of a course of action in tort caused by negligent design or construction was the date when the damage came into existence and not the date when the damage was discovered or could with reasonable diligence have been discovered.

The amendments created by the Act apply only to claims in the tort of negligence and they do not apply to claims made for breach of contract.

Under the Act claims can be brought in tort within three years of the date when the plaintiff had both the knowledge of the damage and the right to bring an action. There is a long stop of 15 years from the time of the act of negligence which caused the damage.

14.3 Limitations and exclusions in MF/1

In MF/1 the conspicuous limitations of liability are those set out in clause 44 under the heading 'Limitations of Liability'. But these are mainly limitations on financial liability and even as such they are not the only limitations on financial liability in the contract.

Financial limitations

Other clauses imposing financial limitations are:

- clause 8.2 – failure to provide bond or guarantee
- clause 16.1 – errors in drawings, etc supplied by the contractor
- clause 34.1 – damages for delay in completion
- clause 34.2 – prolonged delay
- clause 35.8 – damages for failure to pass performance tests
- clause 36.9 – limitation of liability for defects
- clause 43.5 – loss or damage to the works after responsibility has passed to the purchaser.

Time-related limitations

Time-related limitations apply principally to defects as set out in clause 36.9 (limitation of liability for defects) and clause 36.10 (latent defects).

But other time limitations can be found in the contract as follows:

- clause 2.6 – disputing engineer's decisions, instructions and orders
- clause 39.12 – effect of final certificate of payment
- clause 41.1 – notification of claims
- clause 52.1 – notice of arbitration.

Clause 52, the arbitration clause, can also be seen as an exclusion clause to the extent that it excludes litigation in preference for arbitration.

Clause 44.1 – mitigation of loss

In all cases the party establishing or alleging a breach of contract or a right to be indemnified in accordance with the Contract shall be under a duty to take all necessary measures to mitigate the loss which has occurred provided that he can do so without unreasonable inconvenience or cost.

This clause is placed under the heading of limitation of liability but it is more a statement of the general legal position relating to the recovery of damages.

Clause 44.2 – indirect or consequential damage

> Except as expressly provided in Sub-Clauses 34.1 (Delay in Completion) and 35.8 (Consequences of Failure to Pass Performance Tests) for the payment or deduction of liquidated damages for delay or failure to achieve performance and except for those provisions of the Conditions whereby under Sub-Clause 41.2 (Allowance for Profit on Claims) the Contractor is expressly stated to be entitled to receive profit, neither the Contractor nor the Purchaser shall be liable to the other by way of indemnity or by reason of any breach of the Contract or of statutory duty or by reason of tort (including but not limited to negligence) for any loss of profit, loss of use, loss of production, loss of contracts or for any financial or economic loss or for any indirect or consequential damage what-soever that may be suffered by the other.

Many of the items listed here would be recoverable as damages under ordinary principles in a breach of contract or negligence claim. But the intention of the clause is clearly to limit the financial liability of the parties to each other to either express contractual entitlements or direct cost.

The clause could possibly be challenged under the Unfair Contract Terms Act and then it would depend on the facts of the particular case whether or not it survived the test of reasonableness under the Act. But, in any event the clause may not be quite as restrictive as its wording suggests. The question is – what is meant by indirect or consequential damage?

In the case of *Croudace Construction Ltd* v. *Cawoods Concrete Products Ltd* (1978) the contract for the supply of concrete blocks by Cawoods stated that:

> 'We are not under any circumstances to be liable for any consequential loss or damages caused or arising by reason of late supply or any fault, failure or defect in any materials or goods supplied by us or by reason of the same not being of the quality or specification ordered or by reason of any other matter whatsoever.'

Croudace, the main contractors for a school building claimed damages for breach of contract, alleging late delivery and defects. The claim included items for loss of productivity, additional costs of delay in executing the main contract works, and an indemnity against a claim by another sub-contractor for delay to his programme.

It was held that all the losses directly and naturally resulted in the ordinary course of events from Cawoods' alleged breach and were not excluded as 'consequential loss or damage'.

Clause 44.3 – limitation of the contractor's liability

In no circumstances whatsoever shall the liability of the Contractor to the Purchaser under the Conditions for any one act or default exceed the sum stated in the Appendix or if no sum is so stated, the Contract Price. The Contractor shall have no liability to the Purchaser for or in respect or in consequence of any loss or damage to the Purchaser's property which shall occur after the expiration of the Defect Liability Period except as stated in Sub-Clause 36.10 (Latent Defects).

This clause, unlike clauses 44.1 and 44.2, is intended to operate solely as limitation of the contractor's liability. It contains both financial and time-related limitations.

The financial limitations stated here are apparently separate to those stated in the appendix of the contract on damages for late completion but they could be seen, nevertheless, as providing an overall ceiling.

The most obvious situation where the limitation in clause 44.3 might come into effect is when there is failure to pass performance tests. Less obvious is the question of whether clause 44.3 applies in the event of contractor's default which has led to termination of the contract. The provisions of clause 49.3 (payment after termination) suggest that the contractor is liable for the difference between the 'Cost of Completion' and the amount that would have been payable to the contractor for completion. There is no mention of any limit. But clearly if the contractor's price is ridiculously low and he, in effect, abandons the contract because of this then his liability as stated in clause 49.3 might well exceed the limitation of liability as stated in clause 44.3.

This raises a further question. If clause 44.3 does rule over clause 49.3 what is the position if the termination notice specifies more than one default? Is the limit of the contractor's liability then a multiple of the contract price.

Clause 44.4 – exclusive remedies

The Purchaser and the Contractor intend that their respective rights, obligations and liabilities as provided for in the Conditions shall be exhaustive of the rights, obligations and liabilities of each of them to the other arising out of, under or in connection with the Contract or the Works, whether such rights, obligations and liabilities arise in respect or in consequence of a breach of contract or of statutory duty or a tortious or negligent act or omission which gives rise to a remedy at common law. Accordingly, except as expressly provided for in the Conditions, neither party shall be obligated or liable to the other in respect of any damages or losses suffered by the other which arise out of, under or in connection with the Contract or the Works, whether by reason or in consequence of any breach of contract or of statutory duty or tortious or negligent act or omission.

This is both a limitation of liability and an exclusion clause. It limits remedies to those stated in the contract and it expressly excludes all other remedies.

This is certainly one of the most important provisions in MF/1 because it appears to exclude all rights, obligations and liability arising from implied terms as well as excluding actions for negligence.

The consequences of this are mentioned many times through this book and the point which repeatedly arises is this. If the only rights of the parties are express rights – what remedy has the contractor for a breach of contract where there is no express entitlement to payment for the breach? For example, failure by the engineer to carry out his duties.

It could perhaps be argued that breach of an express obligation creates a right which is itself the basis of entitlement whether or not the right is expressly stated. But that is not far short of arguing for an implied term.

Although clause 44.4 is probably more restrictive of the contractor's ordinary legal rights than the purchaser's it is fairly common practice for clause 44.4 to be struck out of the contract by purchasers who are nervous of the rights they might be losing by its inclusion. One unexpected consequence for the purchaser who does this is that he may then find himself liable for a far wider range of financial claims from the contractor.

Clause 36.9 – limitation of liability for defects

The Contractor's liability under this Clause shall be in lieu of any condition or warranty implied by law as to the quality or fitness for any particular purpose or the workmanship of any part of the Works taken over under Clause 29 (Taking-Over) and, save as in this Sub-Clause and in Sub-Clause 36.10 (Latent Defects) expressed, neither the Contractor nor his Sub-Contractors, their respective servants or agents shall be liable, whether in contract, in tort (including but not limited to negligence) or by reason of breach of statutory duty or otherwise, in respect of defects in or damage to such part, or for any damage or loss of whatsoever kind attributable to such defects or damage or any work done or service or advice rendered in connection therewith.

For the purposes of this Sub-Clause the Contractor contracts on his own behalf and on behalf of and as trustee for his Sub-Contractors, servants and agents. Nothing in this Clause shall affect the liability of the Contractor under these Conditions in respect of any part of the Works not yet taken over or his liability for death or personal injury caused by his wilful or negligent acts or omissions.

The reference to 'this Clause' is to the whole of clause 36. In the earlier parts of the clause the contractor's responsibility is detailed as making good defects which appear during the defects liability period (and extensions to the period which occur in respect of removals, etc).

The effect of clause 36.9 therefore is to limit the contractor's responsibility for defects to those defects appearing during the defects liability period (or any extension of it). Consequently the contractor has no ongoing responsibility for latent defects which would allow actions to be brought in contract or in tort during the appropriate limitation periods.

Note that the limitation of liability in the clause purports to extend to negligence and other torts as well as to breach of contract. This, as discussed earlier, needs to be considered in the light of the Unfair Contract Terms Act 1977.

As to the final section of the clause which relates to sub-contractors and servants the intention is to limit the liability of sub-contractors in similar manner to that of the main contractor. A similar provision of more general effect is found in clause 53.1 (sub-contractors, servants and agents). These provisions, however, are of doubtful effect since a sub-contractor sued for negligence would have difficulty raising as his defence the terms of a contract to which he was not a party.

The point came up in the case of *Southern Water Authority* v. *Lewis and Duvivier* (1984) – one of the few cases to reach the courts on the old Model Form 'A'. It was held that the sub-contractors in the case could not obtain the benefit of the exclusion of liability in clause 30(vi) of MF 'A' – a clause virtually identical to clause 36.9 of MF/1 – since the sub-contractors were not signatories to the main contract.

Clause 36.10 – latent defects

If any defect of the kind referred to in Sub-Clause 36.2 (Making Good Defects) shall appear in any part of the Works within a period of three years after the date of the taking-over of such part of the Works the same shall be made good by the Contractor by repair or replacement at the Contractor's option provided that the defect was caused by the gross misconduct of the Contractor as defined below and would not have been disclosed by a reasonable examination prior to the expiry of the Defects Liability Period.

'Gross misconduct' does not comprise each and every lack of care or skill but means an act or omission on the part of the Contractor which implies either a failure to pay due regard to the serious consequences which a conscientious and responsible contractor would normally foresee as likely to ensue or a wilful disregard of any consequences of such act or omission.

This clause provides a minor exception to the exclusion of liability for latent defects in clause 36.9 in that it makes the contractor responsible for three years after taking-over for the very limited category of defects which are:

- caused by gross misconduct – which by definition in the clause seems to mean gross negligence or concealment, and
- could not be found by reasonable examination at the end of the defects liability period.

Note that the defect has to be both caused by gross misconduct and to be non-discoverable. These would be very difficult matters for the purchaser to prove.

There is little legal authority on the matter of gross misconduct but in a building case *William Hill Organisation Ltd* v. *Bernard Sunley & Sons* (1982) it was said of a similar matter – fraudulent concealment – that the question to be asked was were the facts such that the conscience of the contractor should have been affected that it was unconscionable to proceed with the work or so to cover up the defect without putting it right.

14.4 Limitations and exclusions in MF/2

MF/2 contains broadly the same limitations and exclusions as MF/1.

Clauses 32.1 to 32.4 headed 'Limitations of Liability' are the same as clauses 44.1 to 44.4 of MF/1. Clauses 25.7 (limitation of liability for defects) and 25.8 (latent defects) are the same as clauses 36.9 and 36.10 of MF/1.

However, MF/2 contains an additional limitation in clause 25.2 (making good defects) to the effect that replacement of a defective part shall constitute fulfilment of the contractor's obligations for making good defects. In other words the contractor is not obliged to fit the replacement part.

Other differences between MF/2 and MF/1 are that MF/2 does not have secondary limitation clauses in respect of failure to pass performance tests and loss or damage to the works after taking-over.

14.5 Limitations and exclusions in MF/3

The important provisions in MF/3 on limitations and exclusions are:

- clause 5.2   – consequences of rejection
- clause 7.5   – liability for delay
- clause 9.3   – liability for damage or loss in transit
- clause 11.2 – defects liability
- clause 14.1 – liability for accidents and damage
- clause 15.1 – completion.

Clause 5.2 – consequences of rejection

> When goods have been rejected, either under Clause 4 (Tests) or Sub-Clause 5.1 (Notice of Rejection), the Purchaser shall be entitled, provided he does so without undue delay, to replace the goods rejected. There shall be deducted from the Contract Price that part thereof which is properly apportionable to the goods rejected. The Vendor shall pay to the Purchaser any sum by which the expenditure reasonably incurred by the Purchaser in replacing the rejected goods exceeds the sum deducted. All goods obtained by the Purchaser to replace rejected goods shall comply with the Contract and shall be obtained at reasonable prices and, when reasonably practicable, under competitive conditions. Where goods have been rejected as aforesaid the Vendor shall not be under any liability to the Purchaser except as provided in this clause and as may arise under Clause 7 (Time for Delivery).

The effect of this clause to limit the vendor's liability for rejected goods to their replacement cost. That is to say, all consequential loss claims are excluded.

Clause 7.5 – liability for delay

> The Purchaser's remedies under Sub-Clauses 7.3 (Delay in Delivery) and 7.4 (Prolonged Delay) shall be in lieu of any other remedy in respect of the Vendor's failure to deliver goods in accordance with the Contract within the time fixed thereby or any extension thereof or, if no time be fixed, within a reasonable time.

This accords with the basic legal principle that liquidated damages are an exhaustive and exclusive remedy for the matter to which they apply. See the case of *Temloc Ltd* v. *Errill Properties Ltd* (1987) discussed in Chapter 11.

Clause 9.3 – liability for damage or loss in transit

> The liability imposed on the Vendor in this clause shall be accepted by the Purchaser in substitution for all or any other liability on the part of the Vendor arising from the delivery of goods damaged in transit or the non-delivery of goods in consequence of loss in transit.

Clauses 9.1 and 9.2 require the vendor to replace goods damaged or lost in transit free of charge.

Clause 9.3 is a limitation that such replacement is the vendor's sole liability. As with clause 5.2 claims for consequential loss are excluded.

Clause 11.2 – defects liability

The Vendor's liability under this clause or under Clause 5 (Rejection and Replacement) shall be accepted by the Purchaser in lieu of any warranty or condition implied by law as to the quality or fitness for any particular purpose of the goods and save as provided in this clause the Vendor shall not be under any liability to the Purchaser (whether in contract, tort, breach of statutory duty or otherwise) for any defects in the goods or for any damage, loss, death or injury (other than death or personal injury caused by the negligence of the Vendor as defined in Section 1 of the Unfair Contract Terms Act, 1977 or claims in respect of death or personal injury arising under Part 1 of the Consumer Protection Act, 1987) resulting from such defects or from any work done in connection therewith.

This clause corresponds with the provision of clauses 36.9 of MF/1 and 25.7 of MF/2 in limiting the vendor's liability for defects to those appearing within 12 months of delivery (taking-over in MF/1).

However, this clause in its reference to the Unfair Contract Terms Act recognizes that it is not legally possible to exclude negligence claims for death or personal injury.

Clause 14.1 – liability for accidents and damage

If the Vendor, its agents or sub-contractors are on the Purchaser's premises for the purposes of the Contract then, notwithstanding the provisions of Clause 11 (Defects After Delivery), the Vendor shall indemnify the Purchaser against direct damage or injury to the Purchaser's property or person or that of others occurring while the Vendor is working on the Purchaser's premises to the extent caused by the negligence of the Vendor or that of its agents or sub-contractors, but not otherwise, by making good such damage to property or compensating personal injury. Provided that:

(a) the Vendor's total liability for damage to the Purchaser's property [including damage caused by the Vendor's breach of contract, tort (including but not limited to negligence) or breach of statutory duty] shall not exceed £1 million or the Contract Price whichever sum is the greater, and

(b) the Vendor shall not be liable to the Purchaser for any loss of profit or of contracts, or, save as aforesaid, for any loss or damage of any kind whatsoever and whether caused by breach of contract, tort (including but not limited to negligence) or breach of statutory duty of the Vendor or of its agents or sub-contractors or otherwise howsoever.

Under this clause the vendor is liable for:

- direct damage to the purchaser's property

- injury to the purchaser's person
- injury to other persons

whilst working on the purchaser's premises providing the damage or injury is caused by negligence.

However, the limitation in the clause puts a ceiling of £1 million or the contract price (whichever is the greater) on the total of any damage claims. Note that this limitation does not apply to injury claims.

As with other limitations of liability in MF/3 claims for consequential loss are expressly excluded.

Clause 15.1 – completion

> Upon expiry of the defects liability period specified in Clause 11 (Defects After Delivery) the Contract shall be deemed to have been completed and the Vendor shall be under no further obligation or liability whatsoever to the Purchaser whether in contract or in tort (including but not limited to negligence), breach of statutory duty or otherwise unless within 14 days thereafter the Purchaser shall have given the Vendor written notice of any matter in respect of which the Vendor remains obligated or liable to the Purchaser in which event the obligations or liability of the Vendor shall cease immediately upon the Vendor having dealt with the matters specified in the notice. Nothing in this clause shall affect any matter which is the subject of proceedings commenced, whether by way of arbitration under Clause 18 (Arbitration) or otherwise, and which have not been finally determined, prior to the expiry of the said period of 14 days.

The side note of 'Completion' to clause 15.1 appears at first sight a little misleading since the clause deals with legal termination of liability rather than any practical matter of completion.

But the clause can be seen as a completion clause in the sense that it refers to completion of all contractual obligations – that is, completion of the contract.

By clause 15.1 the obligations of the vendor are completed 12 months after delivery unless the purchaser gives notice within 14 days, or has commenced legal proceedings within 14 days, after the 12 month period of any outstanding matters.

14.6 Limitations and exclusions in G/90

The clauses of G/90 which contain limitations and exclusions are as follows:

- clause 3(ii) – failure to provide security for due performance
- clause 21(iv) – indemnification for loss or damage

- clause 22   – limitations on the contractor's liability
- clause 26   – damages for late completion
- clause 27(vii) – failure to pass performance tests
- clause 30A   – limitation of actions
- clause 31(ix) – final certificate
- clause 31(xii) – liability for defects after final certificate.

Clause 3(ii) – failure to provide security for due performance

The Purchaser may … by notice in writing to the Contractor terminate the Contract forthwith, and the Purchaser shall thereupon not be liable for any claim or demand from the Contractor in respect of anything then already done or furnished, or in respect of any other matter or thing whatsoever, in connection with the Contract, but the Purchaser shall be entitled to be repaid by the Contractor all out-of-pocket expenses properly incurred by the Purchaser incidental to the obtaining of new tenders.

Unlike the corresponding clause 8.2 in MF/1 this clause deals with limitation of the purchaser's liability rather than the contractor's liability in the event of the contract being terminated because of the contractor's failure to provide a bond.

It may be intended that the contractor's liability should be limited to the purchaser's costs of obtaining new tenders but that is not what the clause says.

Clause 21(iv) – indemnification for loss or damage

This clause which is considered further in Chapter 15 on accidents, damage and insurances limits the contractor's liability to loss or damage occurring within three years of the works having been taken over.

There are no financial limits on the contractor's liability for such loss or damage unless an upper limit is entered into the appendix to the contract.

Clause 22 – limitations on the contractor's liability

Subject as provided in Clause 26 (Delay in Completion) for the deduction or payment or allowance of liquidated damages for delay, the Contractor shall not be liable to the Purchaser, whether by way of indemnity or by reason of any breach of the Contract, for

(a) loss of use (whether complete or partial) of the Works suffered by the Purchaser; or
(b) loss of profit suffered by the Purchaser; or
(c) loss of any contract suffered by the Purchaser.

This clause corresponds to clause 44.2 of MF/1 in seeking to exclude claims for indirect or consequential loss from the contractor's liability to the purchaser. However, unlike the clause in MF/1, this clause relates only to the contractor's liability and it does not prevent claims of the specified types being made by the contractor against the purchaser.

Note also that there is no overall limitation of the amount of liability in G/90 of the type found in clause 44.3 of MF/1.

Clause 26 – damages for late completion

This is a conventional liquidated damages clause the full text of which is given in Chapter 11.

Liquidated damages act as a limit on liability by excluding claims for unliquidated damages and as in MF/1 the liquidated damages can themselves be capped at an amount stated in the appendix to the contract.

Clause 27(vii) – failure to pass performance tests

The text of this clause is given in Chapter 13 on performance testing. The comments above on the limiting effect of liquidated damages apply also here.

Clause 30A – limitation of actions

> Save in respect of claims arising under Clause 21(iv)(b), the Purchaser shall not bring any proceedings based on a cause of action arising out of or otherwise in connection with this Contract after a period of four years from whichever is the later of the following dates:
>
> (a) the date of Taking-Over of the relevant section or portion (as the case may be) of the Works, and
> (b) the date such section or portion (as the case may be) is brought into use by the Purchaser for its intended purpose.

It is not clear why clause 30A has a different style of suffix than other clauses in G/90 but the point is not thought to be significant.

The clause is somewhat unusual in that it addresses directly the legal limitation periods which apply to actions brought by the purchaser under the contract. Other contracts as will have been noted above seek to achieve the same effect by less direct means.

As stated earlier in this chapter the legal limitation period for a simple contract under English law is six years and it is generally held that in the case of a contract for supply and erection this time runs from the date of completion (or taking-over). Clause 30A of G/90 cuts this period down to four years except for claims for loss or damage or injury to persons. For such claims ordinary legal limitation periods remain in place.

When clause 30A is considered in association with clause 31(xii) – liability for defects after final certificate – it can be seen that for latent damage claims the purchaser has effectively one year after latent damage liability has expired to commence legal proceedings in respect of any such damage.

Note that clause 30A does not seek to alter the legal limitation periods in which the contractor can bring claims against the purchaser. In theory at least these remain in place. However the effect of clause 31(ix) is that financial claims by the contractor must be commenced within one month after the issue of the final certificate. This amounts to much the same thing as directly changing the legal limitation period.

Clause 31(ix) – final certificate

> A final certificate shall, save in the case of fraud or dishonesty relating to or affecting any matter dealt with in the certificate, be conclusive evidence as to the value of the Works unless any proceedings arising out of the Contract whether under Clause 37 (Arbitration) or otherwise shall have been commenced by either party before the final certificate has been issued or within one month thereafter.

This is a straightforward time-related limitation. However, the provisions in this clause differ from those in the corresponding clause of MF/1 – clause 32.12 – in that:

- in G/90 the final certificate is conclusive only as to the value of the works;
- in G/90 the time for challenging a final certificate is limited to one month – whereas in MF/1 it is three months.

Clause 31(xii) – liability for defects after final certificate

> In addition to his obligations under Clause 30(i) (Defects after Taking-Over), the Contractor shall be responsible for making good or for the cost of making good any defect to which this Sub-Clause applies and which may be discovered in any section or portion of the Works during the period of three years (or, if less, the design life specified or agreed) after whichever, in relation to that section or portion, is the later of the following dates:
>
> (a) the date of Taking-Over, and

(b) the date such section or portion is brought into use by the Purchaser for its intended purpose (but not so as to cause such period of three years to extend beyond four years after the date of Taking-Over).

The defects to which this Sub-Clause applies are defects of the kind referred to in Sub-Clause (i) of Clause 30 (Defects after Taking-Over):

(aa) which would not have been discovered upon a reasonable examination by the Engineer at the date of Taking-Over; and

(bb) which are notified by the Purchaser to the Contractor as soon as reasonably practicable after their discovery for the purpose of allowing the Contractor to inspect the defects discovered; and

(cc) which do not arise from a failure by the Purchaser to ensure that the Works or any section or portion of the Works have been operated and maintained in accordance with the Contractor's instructions and good engineering practice at all times.

By this clause the contractor is liable for latent defects which appear within the later of three years of taking-over or putting into use – subject to a four year long stop from the date of taking-over.

This differs significantly from MF/1 where, under clause 36.10, the contractor's three year latent damage liability only applies if there has been 'gross misconduct'.

14.7 *Making good defects generally*

Provisions in contracts requiring the contractor to make good defects for a specified period after taking-over or completion can usually be seen to have two apparently opposing aspects:

- they oblige the contractor to make good defects for which he is responsible, and
- they entitle the contractor to make good such defects.

The entitlement can be seen as a substantial benefit to the contractor in that if defects do occur during the period of liability the contractor will normally be able to rectify them at less cost than he would have to pay as damages to the purchaser.

If the purchaser takes it on himself to rectify defects within the specified liability period thereby depriving the contractor of his right to make good his own defects, the purchaser will be able to recover as damages not the full amount of his expenditure but only the amount of cost that the contractor would have incurred. Recent legal approval to this principle was given in the case of *Tomkinson* v. *The Church Council of St. Michael* (1990) mentioned in Chapter 15.

14.8 Making good defects in MF/1

The following clauses of MF/1 are considered in this section:

- clause 26.1 – defects before taking-over
- clause 36.1 – defects after taking-over
- clause 36.2 – making good defects
- clause 36.3 – notice of defects
- clause 36.4 – extension of defects liability
- clause 36.5 – delay in remedying defects
- clause 36.6 – removal of defective work
- clause 36.8 – contractor to search.

Clause 36.7 (further tests) is considered in Chapter 9 on inspection and testing and clauses 36.9 (limitation of liability for defects) and 36.10 (latent defects) have been covered in section 14.3 of this chapter.

Clause 26.1 – defects before taking-over

Clause 26.1 applies to any part of the works not taken-over. It has three main provisions:

- the engineer can notify the contractor of defects which are to be remedied;
- the contractor shall, at his own speed and at his own expense, make good the defects so notified;
- if the contractor fails to make good the defects the purchaser may do so at the cost of the contractor.

In the event that the purchaser makes good the defects the clause provides:

- the action must be taken without undue delay;
- the steps taken must be reasonable;
- the replacement plant shall comply with the contract;
- the replacement plant shall be obtained at reasonable prices under competitive conditions;
- the contractor shall be entitled to remove and retain the replaced plant.

The final provision of clause 26.1 to the effect that nothing in the clause affects any claim by the purchaser under clause 34 is simply confirming that if delay is caused in making good defects whether undertaken by the contractor or the purchaser the contractor does not get relief from his liability for damages for late completion.

Clause 26.1 can, to some extent, be seen as duplicating the provisions in clause 23.5 (failure on tests or inspection) and it is understood that the 1995 Revision 3 to MF/1 will recognize this by commencing clause 26.1 with a statement to the effect that clause 26.1 is without prejudice to the purchaser's rights under clause 23.5.

Clause 36.1 – defects after taking-over

> In these conditions the expression 'Defects Liability Period' means the period stated in the Special Conditions as the Defects Liability Period or if no such period is stated, 12 months, calculated from the date of taking-over of the Works under Clause 29 (Taking-Over). Where any Section or part of the Works is taken over separately the Defects Liability Period in relation thereto shall commence on the date of taking-over thereof.

Three major points come out of this clause:

- The parties are free to agree whatever defects liability period they wish but unless such a period is stated in the special conditions a 12 month period will apply
- The defects liability period commences on taking-over.
- The defects liability periods for sections or parts taken over separately commence on the dates they are taken-over.

Clause 36.2 – making good defects

> The Contractor shall be responsible for making good by repair or replacement with all possible speed at his expense any defect in or damage to any part of the Works which may appear or occur during the Defects Liability Period and which arises either:
>
> (a) from any defective materials, workmanship or design, or
> (b) from any act or omission of the Contractor done or omitted during the said period.
>
> The Contractor's obligations under this Clause shall not apply to any defects in designs furnished or specified by the Purchaser or the Engineer in respect of which the Contractor has disclaimed responsibility in accordance with Sub-Clause 13.3 (Contractor's Design), nor to any damage to any part of the Works in consequence thereof.

Most of this clause speaks for itself but a few points to note are:

- making good is to be done at all possible speed – see clause 36.5 which follows for default;

- the contractor is responsible for any damage he does to the plant during the defects liability period;
- defects or damage attributable to design for which the contractor is not responsible are outside the scope of the clause.

Clause 36.3 – notice of defects

> If any such defect shall appear or damage occur the Purchaser or the Engineer shall forthwith inform the Contractor thereof stating in writing the nature of the defect or damage. The provisions of this Clause shall apply to all repairs or replacements carried out by the Contractor to remedy defects and damage as if the said repairs or replacements had been taken over on the date they were completed; however the Defects Liability Period in respect thereof shall not extend beyond two years from the date of taking-over such other period as may be stated in the Special Conditions.

The first sentence of this clause is a simple statement that the contractor is to be notified of any defects or damage.

The second sentence has the effect that the contractor's liability for defects in repairs and replacements runs from the date of the repair or replacement subject to a cut-off in this liability two years after taking-over.

Clause 36.4 – extension of defects liability

> The Defects Liability Period shall be extended by a period equal to the period during which the Works (or that part thereof in which the defect or damage to which this Clause applies has appeared or occurred) cannot be used by reason of that defect or damage.

The intention of this clause is to give the purchaser the full benefit of the defects liability period with the plant in use. Therefore, if the plant is shut down for repairs or replacements for which the contractor is responsible, then the defects liability period is extended accordingly.

Clause 36.5 – delay in remedying defects

> If any such defect or damage be not remedied within a reasonable time, the Purchaser may proceed to do the work at the Contractor's risk and expense provided that he does so in a reasonable manner and notifies the Contractor of his intention so to do. The Costs reasonably incurred by the Purchaser shall be deducted from the Contract Price or be paid by the Contractor to the Purchaser.

This clause ties in with clause 36.2 above in that it expressly covers the situation where the contractor does not proceed with repairs or replacements within a reasonable time.

Clause 36.6 – removal of defective work

The Contractor may with the consent of the Engineer remove from the Site any part of the Works which is defective or damaged. If the nature of the defect or damage is such that repairs cannot be expeditiously carried out on the Site.

The reason for this clause is that the plant is the property of the purchaser and its removal therefore requires consent. Perhaps it would be more appropriate for the purchaser to give such consent rather than the engineer.

Clause 36.8 – contractor to search

The Contractor shall, if required by the Engineer in writing search for the cause of any defect, under the direction of the Engineer. Unless such defect shall be one which the Contractor would otherwise be responsible for making good under Sub-Clause 36.2 (Making Good Defects) the Cost of the work carried out by the Contractor in searching as aforesaid shall be borne by the Purchaser and added to the Contract Price.

This clause is intended to deal with the situation where the plant is not functioning properly but the exact cause is not known.

The contractor is entitled to be paid the cost of all searches unless the defect, when discovered, can be shown to be the responsibility of the contractor. There is no provision in the clause for the purchaser to recover his costs if the defect is so shown to be the contractor's responsibility.

14.9 Making good defects in MF/2

The relevant clauses of MF/2 are:

- clause 20.1 – defects before taking-over
- clause 25.1 – defects after taking-over
- clause 25.2 – making good defects
- clause 25.3 – notice of defects
- clause 25.4 – extension of defects liability
- clause 25.5 – delay in remedying defects.

MF/2 does not contain provisions corresponding to those in clauses 36.6 and 36.8 of MF/1 on removal of the defective work and the contractor's obligation to search for defects. But the contractor, of course, has only limited obligations under MF/2 for the plant when it is on site.

Clause 20.1 – defects before taking-over

Without prejudice to the Purchaser's rights under Sub-Clause 17.5 (Failure on Tests or Inspection) if, in respect of any part of the Plant not yet delivered, the Engineer shall at any time:

(a) decide that any work done or materials used by the Contractor or by any Sub-Contractor is or are defective or not in accordance with the Contract, or that such part is defective or does not fulfil the requirements of the Contract (all such matters being hereinafter in this Clause called 'defects'); and

(b) as soon as reasonably practicable notify the Contractor of the said decision, specifying particulars of the defects alleged and of where the same are alleged to exist or to have occurred, then the Contractor shall with all speed and, except as provided in Sub-Clause 19.5 (Resumption of Manufacture or Delivery) at his own expense, make good the defects so specified. Nothing contained in this Clause shall affect any claim by the Purchaser under Clause 24 (Delay).

Clause 20.1 of MF/2 is not exactly the same as the corresponding clause 26.1 in MF/1.

The opening words of clause 20.1 are additional to those in MF/1 but they are merely intended to confirm that the right of rejection under clause 20.1 is independent of rights of rejection under clause 17.5 for failure by the plant to pass tests or inspections. Perhaps the wording could have been clearer, because the only express right of the purchaser under clause 17.5 is to recover the expenses incurred in retesting.

The other significant change from MF/1 is that the purchaser is not given by clause 20.1 of MF/2 the right to make good defects. But MF/1 of course refers to defects before taking-over (and the plant will be on site) whereas MF/2 refers to defects before delivery.

Clause 25.1 – defects after taking-over

In these Conditions the expression 'Defects Liability Period' means the period stated in the Special Conditions as the Defects Liability Period or if no such period is stated, 18 months, calculated from the date of Delivery. Where any Section or part of the Plant is delivered separately the Defects Liability Period in relation thereto shall commence on the date of Delivery thereof.

The only differences between this clause and clause 36.1 of MF/1 are:

- defects liability periods run from dates of delivery;
- defects liability periods are 18 months (not 12 months) unless otherwise stated.

Clause 25.2 – making good defects

The Contractor shall be responsible for making good by repair or replacement with all possible speed at his expense any defect in or damage to any part of the Plant which may appear or occur during the Defects Liability Period and which arises either:

(a) from any defective materials, workmanship or design, or
(b) from any act or omission of the Contractor done or omitted during the said period.

The Contractor's obligations under this Clause shall not apply to any defects in designs furnished or specified by the Purchaser or the Engineer in respect of which the Contractor has disclaimed responsibility in accordance with Sub-Clause 13.3 (Contractor's Design), nor to any damage to any part of the Plant in consequence thereof. The supply to the Purchaser insured and carriage paid to the address stipulated by the Purchaser of a defective or damaged part of the Plant properly repaired or of a part in replacement thereof shall constitute fulfilment by the Contractor of his obligations under this Sub-Clause in respect of that defective or damaged part. If it is reasonably practicable for a defective or damaged part to be returned to the Contractor and the Contractor shall call for its return the Purchaser shall cause it to be returned to the Contractor at the Contractor's risk and expense.

This clause is the same as clause 36.2 of MF/1 up to the word 'thereof' in the middle of the final paragraph. The sentences which follow in clause 25.2 are not found in MF/1.

It will be seen that these sentences provide for:

- limitation of the contractor's liability to replacement of defective parts;
- return to the contractor of replaced parts.

Clause 25.3 – notice of defects

If any such defect shall appear or damage occur the Purchaser or the Engineer shall forthwith inform the Contractor thereof stating in writing the nature of the defect or damage. The provisions of this Clause shall apply to all repairs or replacements carried out by the Contractor to remedy defects and damage except that the period during which the Contractor's responsibility under Sub-

Clause 25.2 (Making Good Defects) shall subsist shall be either 12 months from the date of replacement or renewal or repair of the damage or the unexpired period of the Defects Liability Period whichever is the later to expire.

As with clause 36.3 of MF/1, the contractor is to be notified forthwith of defects or damage. The wording of clause 25.3 then departs slightly from that in clause 36.3 of MF/1 in that although the contractor's responsibility for replaced parts extends for at least 12 months after replacement there is no statement in MF/2 of any cut-off point for this responsibility.

It may well be intended that this should occur 30 months after the original delivery date but it is possible on the wording of the clause that repeat replacements could carry the contractor's responsibility well past this time.

Clause 25.4 – extension of defects liability

This clause is virtually the same as clause 36.4 of MF/1.

Clause 25.5 – delay in remedying defects

This clause also is the same as its corresponding clause in MF/1 – clause 36.5.

14.10 Making good defects in MF/3

MF/3 deals with making good defects in clauses 5.1 (notice of rejection) and 5.2 (consequences of rejection) and in clause 11.1 (defects after delivery).

Clause 5.1 – notice of rejection

> The Purchaser shall be entitled, by notice in writing given within a reasonable time after delivery, to reject goods delivered which are not in accordance with the Contract.

Clause 5.2 – consequences of rejection

> When goods have been rejected, either under Clause 4 (Tests) or Sub-Clause 5.1 (Notice of Rejection), the Purchaser shall be entitled, provided he does so without undue delay, to replace the goods so rejected. There shall be deducted

from the Contract Price that part thereof which is properly apportionable to the goods rejected.

Note that neither under clause 5.1 nor 5.2 is there any express obligation on the vendor to replace rejected goods.

Clause 11.1 – defects after delivery

If within 12 months after delivery there shall appear in the goods any defect which shall arise under proper use from faulty materials, workmanship or design (other than a design made, furnished, or specified by the Purchaser for which the Vendor has in writing disclaimed responsibility), and the Purchaser shall have given notice thereof in writing to the Vendor, the Vendor shall, provided that the defective goods or defective parts thereof have been returned to the Vendor if he shall have so required, make good the defects either by repair or, at the option of the Vendor, by the supply of a replacement. The Vendor shall refund the cost of carriage on the return of the defective goods or parts and shall deliver any repaired or replacement goods or parts as if Clause 6 (Delivery) applied.

Under this clause the vendor is responsible for defects:

- which appear within 12 months of delivery;
- arise under proper use from faulty materials, workmanship or design;
- and of which the vendor has given notice in writing.

Note that there is no mention of any defects liability period for repaired or replaced goods.

14.11 Making good defects in G/90

The contractor's responsibilities for making good defects in G/90 are covered in clause 24 (defects prior to taking-over) and clause 30 (defects after taking-over).

Clause 24 – defects prior to taking-over

This clause is virtually identical to clause 26.1 (defects before taking-over) in MF/1.

Clause 30 – defects after taking-over

The parts of clause 30 relevant to making good are:

- clause 30(i)   – contractor's responsibility to make good defects
- clause 30(ii)   – notice of defects
- clause 30(iii) – extension of defects liability
- clause 30(iv) – delay in remedying defects.

The corresponding clauses of MF/1 are clauses 36.2 to 36.5. They are broadly similar in effect but there are some distinguishing points of detail.

Clause 30(i) – contractor's responsibility to make good defects

The Contractor shall be responsible for making good with all possible speed any defect in or damage to any portion of the Works which may appear or occur during a period of twelve months after whichever, in relation to that portion, is the later of the following dates:

(a) the date of Taking Over, and
(b) the date such portion is brought into use by the Purchaser for its intended purpose (but not so as to cause such period of 12 months to extend beyond two years after the date of Taking-Over).

and which arises either from defective materials, workmanship or design (other than a design made, furnished or specified by the Purchaser and for which the Contractor has disclaimed responsibility in writing within a reasonable time after receipt of the Purchaser's instructions), or from any act or omission of the Contractor during the said period.

This clause combines some of the provisions found in clauses 36.2 and 36.3 of MF/1.

However it is not wholly clear if the intention of sub-clause (b) is to apply the two year cut-off to replacements (as in MF/1) or simply to delay putting into use.

Clause 30(ii) – notice of defects

If any such defect shall appear or damage occur the Engineer shall inform the Contractor thereof stating in writing the nature of the defect or damage. If the Contractor replaces or renews any portion of the Works, the provisions of this clause shall apply to the portion of the Works so replaced or renewed until the expiration of 12 months from the date of such replacement or renewal.

Unlike clause 36.3 of MF/1 there is no mention in this clause of a cut-off point for the contractor's responsibility after two years. But such a cut-off may be intended by clause 30(i).

Clause 30(iii) – extension of defects liability

> The period of 12 months mentioned in Sub-Clauses (i) and (ii) of this clause shall be extended by a period equal to the period during which the Works or that portion thereof in which a defect or damage to which this clause applies has appeared or occurred and cannot be used by reason of that defect or damage.

This is the same as clause 36.4 of MF/1.

Clause 30(iv) – delay in remedying defects

> If any such defect or damage be not remedied within a reasonable time, the Purchaser may proceed to do the work at the Contractor's risk and expense.

The corresponding provisions of clause 36.5 in MF/1 are:

- the purchaser must notify the contractor of his intentions;
- the purchaser must do the work in a reasonable manner;
- the costs incurred by the purchaser may be deducted from the contract price.

Chapter 15

Accidents, damage and insurances

15.1 Introduction

This chapter is essentially about risk management; about the allocation of risks and the insurance of risks.

These are far from straightforward matters. Risks and liabilities may seem easy enough to apportion on paper but the law reports are full of complex cases which expose the difficulties of drafting contractual provisions which will readily identify the risk carrier when disaster strikes.

The recent case of *The National Trust* v. *Haden Young Ltd* (1993) which concerned the destruction of Uppark House by fire resulting from the negligence of sub-contractors working on the roof is a classic example. The court held, amongst other things, that under a JCT Minor Works form of contract a clause imposing an obligation on the employer to insure in the joint names of the employer and the contractor neither expressly nor by implication included sub-contractors.

Interpretation of the contract is frequently but one issue in a damage dispute. First there is the matter of causation – a matter clouded in many cases by fear of admissions which may invalidate insurance policies. And then, of course, there is the problem that insurance is a speciality subject – with its own curiosities of law and language.

Engineer's duty of care

Given the difficulties it is understandable that engineers to contracts are sometimes inclined to leave insurance matters to others they regard as better equipped to look after the purchaser's interests. This is sound policy to the extent that an engineer should always ensure that the purchaser receives the best professional advice. But in so far that an engineer may have, by the terms of his appointment, a general duty to advise the purchaser on all contractual matters the engineer should not assume that the purchaser has of his own accord recognized and understood the obligations and implications of the contract. To do so is to invite a charge of negligence.

In the case of *William Tomkinson & Sons Ltd* v. *The Church Council of St.*

Michael in the Hamlet (1990), again on a JCT Minor Works contract, it was held that an architect who failed to advise the employer of certain risks and the need to insure against them was in breach of his duty of care.

Risks and insurances in the model forms

The model forms for plant contracts place many risks associated with the construction of the works on the purchaser. They also require the purchaser to indemnify the contractor against certain claims. But unlike some standard forms they deal only with the insurance obligations of the contractor. The purchaser is free to carry his own contractual risks with or without insurance.

Such flexibility may have its advantages but it does emphasize that the purchaser has decisions to make and insurances to consider in respect of each and every contract.

15.2 Risks generally

Standard forms vary in the extent to which they identify and deal with particular risks. This is inevitable in that a contract for, say, the supply of goods under MF/3 involves by its very nature less risks than a contract for supply and erection of plant under MF/1. But even where the risks are similar the coverage in contracts is not consistent. For example many standard forms, including the IChemE forms, make it very clear that the purchaser is responsible for his own property but the model forms for plant contracts do not directly address the matter.

Risks in plant contracts

The risks in a typical contract for the supply and erection of plant include:

- damage to the plant prior to taking-over
- damage to the plant after taking-over
- faulty materials and workmanship
- thefts and vandalism
- design defects
- damage to the purchaser's property
- damage to third party property
- consequential losses from damage
- injuries to the contractor's employees
- injuries to the purchaser's employees
- injuries to third parties.

Allocation of risks

The policy which underlies most standard forms is that risks should be allocated to the party best able to control them.

Thus taking-over of the works by the purchaser is usually seen as a watershed in respect of the plant itself. Up to that time the contractor has care of the works and is generally responsible for damage whereas afterwards the purchaser becomes responsible - subject to the proviso that the contractor is responsible for any damage he causes whilst remedying defects.

Responsibility for damage or injury to third parties usually follows the cause but damage to the purchaser's property is the purchaser's risk in some contracts.

The contractor is almost invariably responsible for the quality of work and carries the risks of faulty workmanship and materials. Responsibility for defective design would appear to fall naturally on the party which undertook the design but that is not always the case.

Excepted risks

Excepted risks, or the purchaser's risks as they are called in the model forms, are those risks which are expressly excluded from the contractor's responsibility. Typically they include:

- acts or omissions of the engineer, purchaser or his servants;
- use or occupation of the works;
- damage which is the inevitable or unavoidable consequence of the construction of the works;
- war, riots and similar non-insurable events.

Broadly excepted risks fall into three categories:

- fault or negligence of the purchaser;
- matters under the control of the purchaser;
- matters not the fault of either party.

The logic of the first two categories is obvious enough; the argument for the third category, where it applies, is that the purchaser is the party better able to carry the risk.

Limitations on liability

Plant and process contracts, unlike construction contracts usually state limitations on the liability of the contractor to the purchaser for his acts and defaults. In MF/1 for example there is an overall financial limit (often the contract price) and an exclusion of liability for consequential loss.

Such limitations, however, apply only between the contractor and the purchaser and they do not protect the contractor against third party claims.

15.3 Insurances generally

Certain insurances are required by law, for example, motor insurances and employer's liability. In addition contracts may impose insurance requirements on one or both parties to ensure that funds are available to meet claims and to facilitate the completion of the works.

The model forms for plant contracts specify only the insurances which the contractor must carry. The IChemE forms, in contrast, place obligations to insure on both parties.

Common provisions

The common provisions of plant, process and construction contracts are:

- the contractor is responsible for care of the works until completion;
- the contractor must insure the works to their full replacement cost;
- the contractor must indemnify the purchaser against claims for injury to persons or damage to property;
- the contractor must insure against that liability.

Other provisions

According to the amount of detail in the insurance clauses of the contracts other provisions may cover:

- approval of insurers
- production of documentary evidence
- minimum levels of cover
- maximum levels of excess
- the purchaser's rights if the contractor fails to insure
- professional indemnity
- joint insurances.

Professional indemnity for consultants

Purchasers who engage consulting engineers as designers almost invariably require that they have professional indemnity insurance.

Where the contractor is responsible for design, either in-house or through consulting engineers, it might appear on the face of it to be no concern of the purchaser's whether or not professional indemnity insurance is maintained. However, it is not always seen that way and it is not uncommon for such insurance to be required. For cover where consulting engineers are the designers there are two problems.

Firstly, there is the point that professional indemnity insurance is usually on a claims made basis so that the cover is only effective in respect of the year in which the claim is made. Thus once a policy lapses there is no cover for past work. Consequently a contractual requirement for such insurance needs to be drafted to ensure that cover is maintained for the legal limitation period rather than merely the construction period as for other insurances. That raises the question – how realistic is it to monitor the maintenance of such insurance over a protracted period of years.

Secondly, there is the problem that the legal responsibility of a professional designer is limited at common law to the exercise of reasonable skill and care and his professional indemnity cover is similarly limited. Therefore a claim on a fitness for purpose basis will have no access to such insurance.

Contractor's in-house design insurance

Contractors who undertake in-house design can insure against the negligence of their own designers. The cover is usually defined as being in respect of a negligent act, error or omission of the contractor in performance of his professional activities. But see the *Wimpey* case below.

The need for such insurance arises because a contractor's all risk policy usually excludes design entirely or limits the indemnity to damage caused by negligent design to third party property or construction works other than those designed.

An ordinary professional indemnity policy does not cover the contractor against the problem of discovery of a design fault before completion. At that stage there is no claim against the contractor as there would be against an independent designer. To overcome this, contractors usually seek a policy extension giving first party cover. In effect this amounts to giving the construction department of the contractor's organization a notional claim against the design department.

The complexities of in-house design claims were revealed in the case of *Wimpey Ltd* v. *Poole* (1984) which concerned a contract under the ICE Fourth edition modified for contractor's design. During the construction of a new quay wall at Vospers Southampton shipyard movement of the quay wall occurred and extensive remedial works were necessary. The cause was found to be errors by Wimpey's in-house designers in their assumptions on soil mechanics. Wimpey sought to recover the cost of the remedial works from their insurers and set out to prove their own negligence even to the point of advancing the argument that a company of their standing should be judged by a more stringent and exacting task than an ordinary practitioner.

Wimpey failed to establish negligence although they did win the argument that the words 'negligent act, error or omission' should be construed to include any error or omission without negligence.

Terminology

Insurance clauses in contracts often use phrases which are not particularly clear in themselves but which have particular meanings to insurers. These are a few such phrases.

● Subrogation

This is the legal right of an insurer who has paid out on a policy to bring actions in the name of the insured against third parties responsible for the loss.

● All risks

An all risks policy does not actually cover all risks since invariably there will be exceptions. However the effect of an all risks policy is to place on the insurer the burden of proving that the loss was caused by a risk specifically excluded from cover. In contrast, under a policy for a specified risk it is the insured who must prove that his loss was caused by the specified risk.

● Joint names

Insurance in joint names provides both parties with rights of claim under the policy and it prevents the insurer exercising his rights of subrogation one against the other.

● Cross liability

The effect of a cross liability provision in a policy is that either party can act individually in respect of a claim notwithstanding the policy being in joint names.

15.4 Accidents, damage and insurances in MF/1

The provisions of MF/1 relating to accidents, damages and insurances fall under five headings:

- accidents and damage   – clause 43
- limitations of liability   – clause 44
- purchaser's risks   – clause 45
- *force majeure*   – clause 46
- insurances   – clauses 47 and 48

Plus:

- extraordinary traffic   – clause 21.1
- special loads   – clause 21.2
- extraordinary traffic claims – clause 21.3
- waterborne traffic   – clause 21.4

In summary these provisions are as follows.

Accidents and damage

Clause 43.1 – care of the works
The contractor is responsible for care of the works until the date of taking-over and care of any outstanding work until completion. In the event of termination, responsibility for care of the works passes to the purchaser.

Clause 43.2 – making good loss or damage to the works
The contractor is to make good any loss or damage to the works except to the extent it is caused by the purchaser's risks.

Clause 43.3 – damage to works caused by purchaser's risks
The purchaser has six months from the happening of an event to require the contractor to make good loss or damage caused by the purchaser's risks either at a price to be agreed or to be determined in arbitration.

The clause itself does not consider the consequences of any delay or failure by the purchaser in stating his requirements. For delay, an extension of time would certainly be due under clause 33.1 but the position is less clear in respect of claims for money. As to failure, cir-cumstances can be envisaged where it is not possible for the contractor to complete the works until the purchaser states his requirements. Termi-nation might then follow – but whether that would be termination for purchaser's breach or termination for *force majeure* would depend on the circumstances.

A commonly arising problem when there is loss or damage to the works is whether or not the responsibility lies with the contractor or the pur-chaser. That dispute may take some time to resolve and in the meantime delay builds up. Usually the contractor will decide to proceed with the making good whilst reserving his position on liability.

Clause 43.4 – injury to persons and property whilst contractor has responsibility for care of the works

The contractor is liable for and is to indemnify the purchaser in respect of all personal injury and damage to property claims which arise out of the execution of the works whilst the contractor is responsible for their care.

However, the contractor is not liable for, and the purchaser is to indemnify the contractor against, claims which result from acts or neglects of the purchaser and other contractors; and also claims for damage to property which are the inevitable consequence of the execution of the works.

Note that the exclusion of the contractor's liability on the grounds of 'inevitable consequence' does not apply to personal injury claims.

It is not absolutely clear if the purchaser's premises in which the works are being installed come within the scope of this clause. The clause reads as though it applies to third party claims. But it does use the phrase 'any property' which might be extendable to the purchaser's premises. See clause 44.3 below which implies the contractor is liable for damage to the purchaser's property prior to the end of the defects liability period.

Clause 43.5 – injury to persons and damage after responsibility for care of works passes to purchaser

After the works are taken over the contractor remains liable for and is to indemnify the purchaser in respect of claims for personal injury or damage to property (other than the works) resulting from the contractor's negligence, defective design, materials or workmanship.

Loss or damage to the works is covered by the provisions of clause 36 (defects liability)

Clause 43.6 – accidents or injury to workmen

The contractor is to indemnify the purchaser against personal injury claims in respect of the contractor's or sub-contractor's employees. However the indemnity does not apply to the extent the claim results from an act or default of the purchaser or another contractor. The purchaser is to indemnify the contractor against such claims.

Clause 43.7 – claims in respect of damage to persons or property

The contractor is entitled to take-over the conduct of any proceedings brought against the purchaser for which the contractor is liable.

The purchaser must afford the contractor all available assistance but is entitled to security for costs and reimbursement of costs involved.

Limitations of liability

Clause 44.1 – mitigation of loss
The party being indemnified has a duty to mitigate his loss providing he can do so without unreasonable inconvenience or cost. This reflects the common law position.

Clause 44.2 – indirect or consequential damage
Neither party is entitled to claim against the other for breach of contract or negligence for loss of profit or other claims for economic loss or indirect or consequential damage except as expressly provided for in the contract.

It is open to argument whether this clause is compatible with the Unfair Contract Terms Act 1977. See the comments in Chapter 14.

Clause 44.3 – limitation of contractor's liability
The liability of the contractor to the purchaser for 'any one act or default' is the sum stated in the appendix or if no sum is stated the contract price. The contractor is not liable to the purchaser for damage to the purchaser's property after the expiration of the defects liability period.

Clearly this clause cannot limit the contractor's liability in respect of direct third party claims and it is presumably only intended to apply to the purchaser's own claims against the contractor. It is not thought to apply to indemnities but if it can be argued that it does it leaves the purchaser seriously exposed.

Clause 44.4 – exclusive remedies
The rights, obligations and liabilities of the parties as set out in the contract are exhaustive of their rights against each other whether in contract or in tort. For further comment on this clause see Chapter 14.

Purchaser's risks

Clause 45.1 – purchaser's risks
These can be summarized as:

- fault, error, etc in the design for which the engineer or the purchaser is responsible;

- use or occupation of the site by the works;
- use or occupation for the purposes of the contract;
- interference which is the inevitable result of the construction of the works;
- damage which is the inevitable result of the construction of the works;
- use of the works by the purchaser;
- act, neglect or omission of the engineer, purchaser or other contractors;
- *force majeure* except to the extent it is covered by the contractor's insurances.

The purchaser's risks as listed above and detailed in clause 45.1 apply only to care of the works and not to third party damage. Nevertheless they are extensive enough to warrant the purchaser including them in his own insurances to the extent he is able to do so.

Force majeure

Clause 46 is covered in detail in Chapter 19. The only comment to add here is on the link between the purchaser's risks and *force majeure*.

The item of concern to the purchaser would be does the phrase 'any circumstances beyond the reasonable control of either of the parties' fall to be considered as purchaser's risks? The answer is 'yes' to the extent that they are not covered by the contractor's insurances. This suggests that the purchaser should take more than a passing interest in the coverage of the contractor's insurances.

Insurance

Clause 47.1 – insurance of works
The contractor must insure the works and contractor's equipment up to their full replacement value in the joint names of the contractor and the purchaser. The insurance is not required to cover the purchaser's risks. The insurance is stated to run from the date of the letter of acceptance until whichever is the earlier of:

- fourteen days after the date of issue of a taking-over certificate, or
- fourteen days after the date when responsibility for care of the works passes to the purchaser.

The wording here is clearly defective since by clause 43.1 (care of the works) responsibility for care of the works passes to the purchaser on the date of issue of a taking-over certificate – that is to say, the two dates in clause 47.1 are the same.

However, it is understood that the 1995 Revision 3 to MF/1 will correct this by amending the second date to apply only to taking-over following termination.

Clause 47.2 – extension of works insurance
The contractor is required to extend 'so far as reasonably possible' insurance cover to the periods when he is on site making good defects or carrying out or supervising testing.

Clause 47.3 – application of insurance monies
All monies recovered under a works insurance policy shall be applied towards the repair of any damage. The clause also states 'this provision shall not affect the Contractor's liabilities under the Contract'. This may be intended to emphasize the point that the contractor's liability is not limited to the amount of any insurance cover.

Clause 47.4 – third party insurance
The contractor is required to take out third party (public liability) insurance to an amount not less than that stated in the special conditions.

Because of difficulties with this type of insurance being in joint names the clause requires the terms of the policy to include a provision that the insurers indemnify the purchaser in respect of claims made directly against the purchaser for which the contractor is entitled to indemnity under the policy.

Clause 47.5 – insurance against accidents, etc to workmen
The contractor is to take out employer's liability insurance and to ensure that the employees of sub-contractors are similarly covered.

Clause 47.6 – general insurance requirements
All insurances are to be:

- with an approved insurer
- on terms approved by the purchaser
- available for inspection.

The contractor is required to provide:

- proof of payment of premiums
- evidence of insurance cover
- details of any policy changes.

Clause 47.7 – exclusions from insurance cover
Insurance policies are not required to cover:

- defective work
- the purchaser's risks
- consequential loss
- fair wear and tear
- motor or other insurances required by law.

Clause 48.1 – remedy on failure to insure
If the contractor fails to insure the purchaser can do so and deduct the premiums from monies due to the contractor.

Clause 48.2 – joint insurances
This clause ensures that an insurer having paid out under a claim does not have rights of subrogation against any other insured party under the policy.

Extraordinary traffic and special loads

Clause 21.1 – extraordinary traffic
The contractor is required to use every reasonable means to prevent damage to roads and bridges on route to the site.

Clause 21.2 – special loads
If the contractor considers that moving plant or contractor's equipment is likely to cause damage he must notify the engineer of his proposals for protecting or strengthening. Unless the engineer directs within 14 days that such protecting and strengthening is unnecessary the contractor can carry out his proposals. The contractor is entitled to have the cost of such work added to the contract price. In practice, of course, it will rarely be the contractor who carries out any necessary protecting or strengthening – that will be undertaken by the highway authority or similar. Nevertheless, the purchaser is still liable for the bills and this is a matter which requires careful attention by the engineer.

The matter of delays which may result from protecting and strengthening works is not dealt with directly in MF/1 but the contractor could claim an extension of time under clause 33.1 on the grounds of circumstances beyond the contractor's control.

Clause 21.3 – extraordinary traffic claims

The contractor is to report any claims he receives for damage to roads or bridges to the engineer. The purchaser is to negotiate and pay all sums due and to indemnify the contractor in respect of such claims.

If the claims result from negligence of the contractor or failure by the contractor to perform his obligations of using reasonable means to avoid damage or failing to give notice of likely damage then the engineer is required to certify the amount to be deducted from the contract price.

What this latter clause presumably means is that the engineer carries out some apportionment of the costs incurred. And, on the assumption that the purchaser is meeting the total of the costs in bills outside the contract price, that leaves a sum to be deducted from the contract price as monies due from the contractor.

Again, in practice, this does not work in the manner described. Generally, when damage occurs loss adjusters instructed by insurance companies take on the role of apportionment of responsibility and the role of the engineer in the process is limited.

Clause 21.4 – waterborne traffic

This clause merely substitutes waterways for highways.

15.5 Accidents, damage and insurances in MF/2

In MF/2 the provisions relating to accidents, damage and insurances are grouped under three headings:

- accidents and damage – clause 31
- limitations on liability – clause 32
- insurance – clause 33.

There is no need in MF/2 for provisions on purchaser's risks or for requirements on third party insurances because of the physical separation of the parties up to delivery and the clean transfer of responsibility on delivery.

Accidents and damage

Clause 31.1 – care of the plant

The contractor is responsible only for care of the plant until it is delivered. There are no exceptions.

Clause 31.2 – making good loss or damage to the plant
The contractor is required to make good loss or damage to the plant at his own expense.

Limitations on liability

Clauses 32.1 to 32.4 of MF/2 covering:

- mitigation of loss
- indirect or consequential damage
- exclusive remedies

are the same as clauses 44.1 to 44.4 of MF/1.

Insurance

Clause 33.1 – insurance of plant
Unlike MF/1 the insurance of plant requirements do not apply generally to the contractor's responsibility for care of the works but only to plant which has vested in the purchaser. The purchaser's interest in the plant is to be noted on the insurance policy.
The risks to be covered are to be specified in the special conditions.

Clause 33.2 – application of insurance monies
All monies received under the policy to be applied towards making good loss or damage.

Clause 33.3 – general insurance requirements
The same as clause 47.6 of MF/1.

Clause 33.4 – exclusions from insurance cover
The following exclusions are permitted:

- defective work
- consequential loss
- fair wear and tear.

The purpose of this clause is not clear having regard to the provision in clause 33.1 that the risks are to be specified in the special conditions.

Clause 33.5 – remedy on failure to insure
The same as clause 48.1 of MF/1.

15.6 Accidents, damage and insurances in MF/3

The clauses of MF/3 which cover accidents and damage are:

- clause 9.1 – damage in transit
- clause 9.2 – loss in transit
- clause 9.3 – liability for loss in transit
- clause 14.1 – liability for accident and damage.

There are no provisions in respect of insurances.

Damage or loss in transit

Clause 9.1 – damage in transit
The vendor is responsible for the goods to the place of delivery.

Clause 9.2 – loss in transit
The vendor is responsible for goods lost in transit.

Clause 9.3 – liability for damage or loss in transit
The liability of the vendor to repair or replace goods damaged or lost in transit is in substitution for all and any other liability of the vendor arising from damage or loss in transit.

Liability for accidents and damage

Clause 14.1
The vendor is to indemnify the purchaser against damage or injury to the purchaser's property or person occurring whilst the vendor is working on the purchaser's premises but only to the extent that such damage is caused by negligence.

The vendor's liability is not to exceed £1 million or the contract price whichever is the greater. The vendor is not liable for consequential loss.

15.7 Accidents, damage and insurances in G/90

G/90 deals with accidents, damage and insurances as follows:

- clauses 21(i) to 21(iii)   – liability for accidents and damage
- clauses 21(iv) to 21(v)   – indemnification
- clause 22   – limitations on the contractor's liability
- clauses 23(i) to 23(ii)   – insurance of the works
- clauses 23(iii) to 23(iv) – damage to persons and property.

Liability for accidents and damage

Clause 21(i)
The contractor is responsible for care of the works until they are taken over and shall make good any loss or damage from any cause other than the excepted risks at his own cost.

Clause 21(ii)
In the event of loss or damage from the excepted risks the contractor shall make good if required to do so by the purchaser and shall be paid a sum ascertained in like manner to the valuation of variations.

Clause 21(iii)
The excepted risks are:

- war, invasion, hostilities, etc
- civil war, rebellion, etc
- pressure waves, etc
- radioactivity, nuclear hazards, etc
- defect, error or omission in design specified by the purchaser for which the contractor has disclaimed responsibility.

Indemnification

Clause 21(iv)
The contractor is to indemnify the purchaser against claims for damage to persons or property which occur:

- before the works have been taken over and arise from the carrying out of the works or design for which the contractor is responsible. But such indemnity is reduced proportionately to the extent that the damage is caused by the purchaser or is the inevitable result of carrying out the works;
- within three years of the works having been taken over and is caused by the contractor's default or any fault in the design for which the contractor is responsible.

The liability of the contractor to indemnify the purchaser in respect of damage to the purchaser's property is limited to direct physical damage (that is, claims for consequential loss are excluded).

There is no limit on the liability of the contractor to indemnify the purchaser unless upper limits on such liability are entered in the appendix.

Note that limit is separate from the limitations on the contractor's liability to the purchaser in respect of consequential loss.

Clause 21(v)

This is much the same as clause 43.7 of MF/1. In the event of claims being made against the purchaser for which the contractor is liable, the contractor is entitled to take control of the proceedings.

Limitations on the contractor's liability

Clause 22 provides that the contractor is not liable to the purchaser by way of indemnity or for breach of contract for consequential losses listed as:

- loss of use
- loss of profit
- loss of any contract.

Insurance of the works

Clause 23(i)

The contractor is required to insure the plant to full replacement value until it is delivered to site against all risk other than the excepted risks. Where the purchaser has an insurable interest this shall be noted in the policy.

Clause 23(ii)

The contractor is required to insure in joint names all plant delivered to site until taking-over against loss or damage from all causes other than the excepted risks.

All monies received under any policy are to be applied to making good any damage. If the purchaser is responsible for the loss or damage and an excess on the policy is entered in the appendix the purchaser shall bear the costs of repair up to the amount of the excess. But this applies only when the purchaser is responsible and not for other contractors of the purchaser.

Damage to persons and property

Clause 23(iii)

The contractor is required to carry third party insurance for a sum not less than that stated in the appendix throughout the performance of the contract. Such insurance is to be effected with an approved insurer and on terms approved by the purchaser. The policy shall contain a provision that the insurer indemnifies the purchaser against claims for which the contractor is liable.

The insurance does not limit the contractor's obligations and liabilities under the contract.

Clause 23(iv)

The contractor is required to produce the policies for inspection and current premium receipts. The purchaser is to be notified of any changes or endorsements.

If the contractor fails to insure the purchaser may do so and deduct the premiums from monies due to the contractor.

Chapter 16

Variations

16.1 Introduction

Variation clauses are necessary in all contracts where the works or goods as originally defined or specified are subject to change.

The contractor's obligation is to complete the work or supply the goods specified in the contract. In the absence of an express provision in the contract empowering change, the contractor is neither obliged nor entitled to make changes.

On that basis alone it is understandable that variation clauses attract a good deal of attention and scrutiny to see how well they accommodate change. But added to that is the point that in plant, process and construction contracts, variations provide by far the most common cause of adjustment of the contract price.

Purchasers may view variation clauses as something of an obstacle in the path towards certainty of price but few purchasers (less still their engineers) could say with confidence that they would under no circumstances order any changes. So to the extent that variation clauses preserve the integrity of the contract in the face of change and also to the extent that they lay down rules for the valuation of variations such clauses have to be seen as being of benefit to the purchaser rather than to the contractor.

There is obviously some benefit to the contractor in having a defined scheme which regulates the ordering and valuation of variations, particularly if the scheme permits the contractor to propose his own variations. But the contractor would always have a legal right to payment for work or goods ordered by the purchaser whether or not there was a variation clause. That right might well be exercised by a claim outside the contract on *quantum meruit* basis in the absence of a variation clause. And such a claim might try to encompass the original contract work with the aim of putting all of the work on a cost-plus basis. See the case discussed later in this chapter and in Chapter 17 of *McAlpine Humberoak* v. *McDermott International* (1992).

Variations in the model forms

All the model forms of contract covered in this book, MF/1, MF/2, MF/3 and G/90 have variation clauses which include the essentials of empowering the engineer or purchaser to order variations and providing rules for the valuation of variations.

MF/3 and G/90 take their variation clause from the old MF 'A'; whereas MF/1 and MF/2 have an improved version of that clause.

All four contracts place a limit of 15% of the contract price on the net value of variations which can be ordered without the consent of the contractor. In the case of MF/1 and MF/2 any variations beyond this limit also need the consent of the purchaser.

16.2 Variations generally

As discussed above variation clauses are included in contracts principally to permit the purchaser to make change and to order extras. Other matters which are commonly included in such clauses are:

- definition of variations and/or scope of the clause;
- power of the engineer (or purchaser) to order variations;
- procedure for ordering variations;
- requirements on consultation with the contractor;
- contractor's proposals for variations;
- financial limits on the net value of additions or deductions;
- notice and confirmation requirements;
- valuation of variations.

Scope of variation clauses

A broad definition of a variation is work which is not expressly or implicitly included in the original contract price. MF/1 defines its meaning of variation as any alteration of the works whether by way of addition, modification or omission.

However, even with such a definition there can be argument as to whether a variation comes within the scope of the variation clause of the contract. Clearly the varied works need to have some relationship to the original work and need to be of the same character. But it does not follow that they will be a variation simply because they are carried out at the same time as the original works and on the same site. In a case under the ICE Fifth edition Conditions of Contract, *Blue Circle Industries plc* v. *Holland Dredging Co (UK) Ltd* (1987) it was held that work in attempting to form an island bird sanctuary with dredged material was not within the scope of the variation clause.

In that case the decision was perhaps influenced by the requirement in the ICE Conditions that variations should be either necessary for completion or desirable for the satisfactory functioning of the works. There is no similar express requirement in the model forms for plant contracts and it may be going too far to say that such a requirement could be implied. The model forms do seem to permit variations which may be neither strictly necessary nor desirable but may simply be cost saving measures.

Nevertheless, the scope of the variation clauses in the model forms must be constrained if only by the contractor's ability to undertake the work involved. That point is recognized in the model forms which require the contractor to give notice if any variation is likely to prevent or prejudice performance of his obligations.

Omission variations

Variation clauses commonly include the power to make omissions – and all four model forms for plant contracts do so.

However, that does not entitle the purchaser to take work out of the contract and give it to others. To do so amounts to a breach of contract for which the purchaser is liable in damages.

The leading legal case on this, *Carr* v. *Berriman Pty Ltd* (1953) was followed in another Australian case – *Commissioner for Roads* v. *Reed & Stuart Pty Ltd* (1974) where it was held:

- it was a concept basic to the contract that the contractor should have the opportunity of performing the whole of the contract work;
- the variation clause entitled the employer only to omit work from the contract works altogether. It did not permit the taking away of a portion of the contract work for it to be performed by another contractor.

There is some uncertainty on how these rules apply to provisional sums. Much depends upon the wording of the contract in relation to the expenditure of such sums. But usually they can be deleted from the contract price without a variation order and without creating any breach of contract.

Duty to order variations

The model forms for plant contracts use the phrase 'the Engineer shall have the power' in relation to the making of variations. That wording seems to imply the exercise of discretion on the part of the engineer. It contrasts, for example, with the wording in the ICE Conditions 'the Engineer shall order any variation ... necessary for the completion of the Works'.

Clearly under the ICE Conditions, contractors are entitled to a variation order if the specified work or methods prove impossible to perform. Thus in *Yorkshire Water Authority* v. *Sir Alfred McAlpine & Son (Northern) Ltd* (1985) it was held that incorporation of a method statement into a contract made it a specified method of construction and when that method proved impossible the contractor was entitled to a variation for an alternative method of working. And in *Holland Dredging (UK) Ltd* v. *Dredging & Construction Co Ltd* (1987) it was held that the contractor was entitled to a variation order for measures in rectifying a shortfall in backfill to a sea outfall pipe arising from loss of dredged material.

However, even where the wording of the contract seems to suggest discretion, for example by use of the phrase 'the Engineer may order' or by wording as found in the model forms for plant the proper interpretation of the contract is probably that there is a duty on the engineer to order variations which are necessary for completion of the works. But that, of course, is subject to consideration of the contractor's responsibility for design and impossibility of the contractor's own making or default.

Payment for variations

In many contracts it is made a condition precedent to the contractor's right to payment for variations that instructions should have been given. In such contracts the absence of written orders or instructions can leave the contractor without any right to payment under the contract even though the purchaser has received the benefit of the extra work.

The variation clauses of MF/3 and G/90 both commence with a statement that the contractor shall not alter any of the works except as directed in writing. And later words in those clauses further suggest that a direction in writing is required before there can be any adjustment of the contract price.

The wording of the variation clauses in MF/1 and MF/2 is not as clear. But taken with other provisions of those contracts relating to instructions in writing the position is probably the same in that if the contractor proceeds with variations which are not properly ordered he risks his rights to payment.

16.3 Limitations on the value of variations

Various model forms place a financial limit, expressed in terms of a percentage of the contract price, on the value of variations which can be ordered without the consent of the contractor.

In the model forms for plant contracts this limit is 15%. In the IChemE model forms the limit is 25%.

The principal effect of these limits is that the contractor is not obliged to undertake variations which would take the revised contract price outside the stated percentages. Without such limits the contractor under the wording of most standard forms would be in breach of contract in refusing to perform any properly ordered variations whatever their value.

With limits, however, the contractor has a choice. He can either accept variations which go beyond the limits and undertake them as part of the contract works or he can exercise his right of refusal.

Problems of limits

Reasonable as this may be from a contractual viewpoint it does raise some practical problems:

- what is the position if there is a disagreement on the value of variations?
- how are the works to be completed if variations exceeding the limits are necessary for completion?
- what happens when an omission variation outside the limits is ordered?
- do variations proposed by the contractor fall within the scope of the limits?

The answers to these questions will sometimes be found in the variation clauses themselves – the difficulty of the point having been anticipated. In other cases it is a matter of applying general legal principles to arrive at the intention of the parties.

Disagreements on value

For disagreements on value the IChemE model forms refer the matter to an 'Expert' for determination. The model forms for plant contracts, lacking such a procedure, appear to rely on arbitration – although this could involve in some cases progress of the works being suspended pending the outcome.

Essential variations

None of the model forms deals directly with the problem of it being necessary to exceed the limits in order to achieve completion. Perhaps it can be argued that the purchaser warrants that the works can be completed within the limits and he must take the consequences if they cannot. The contractor is certainly in a strong position in these circumstances, since he can apparently demand whatever price he wants to complete with the threat of leaving the works uncompleted once work up to the value of the limits has been executed.

Excessive omission variations

With omission variations outside the limits the position is probably that the contractor is entitled to be paid as a minimum the contract price less the stipulated maximum percentage deduction. Thus again the contractor appears to have a choice – to either comply with the variation whilst demanding the minimum contract price or to reject the variation and to perform work up to the minimum contract price.

Contractor proposed variations

As to variations proposed by the contractor the wording of the contract may indicate whether the value of such variations is to be included in the amounts governed by limits. The IChemE model forms are reasonably clear on this and the value is included. The model forms for plant contracts are less clear and it is arguable that they are not included. It would certainly be an odd situation if the contractor was able to propose a variation which because of its value could have the effect of relieving him of completing the works.

However, perhaps it is against this very possibility that MF/1 and MF/2 have the requirement that the purchaser as well as the contractor must consent to variations which exceed the stipulated limits.

Limitations in remeasurement contracts

Most plant contracts are lump sum and the question of whether changes in quantities are to be taken into account in the amounts governed by limits does not apply. However Supplement No. 1 to MF/1 issued in 1993 is concerned with payments which are to be determined in whole or in part by remeasurement.

The complications which can arise when a contract contains both remeasurement provisions and limitation of variation provisions are considerable. Particularly so if there is any suggestion in the contract that changes in quantities are to be regarded as variations; For a legal case on this issue see: *Arcos Industries Pty Ltd* v. *The Electricity Commission of New South Wales* (1973).

16.4 *Valuation of variations*

The rules for the valuation of variations are usually found in the contract itself. All the standard forms for plant, process and construction contracts have such rules but many sub-contract forms overlook the matter.

In the absence of express rules the contractor (or sub-contractor) is generally entitled to a fair valuation for extra work which is properly ordered. This is likely to be claimed on a cost-plus basis. But see the case of *Laserbore Ltd* v. *Morrison Biggs Wall* (1993) discussed below on what constitutes a fair and reasonable valuation.

Express rules

In lump sum contracts, particularly those involving contractor's design, the current preferred method of valuing variations is to obtain and accept the contractor's quotation prior to the variation being confirmed.

But this has its difficulties; sometimes in respect of timing and sometimes in respect of the amount of the quoted price. Therefore even those contracts which suggest a quotation as the starting point for valuation tend to have rules either to assess the reasonableness of the quotation or to provide alternative methods of valuation.

Thus the scheme in MF/1 and MF/2 is:

- by contractor's quotation; failing which,
- by reference to rates in the schedules of prices if applicable; failing which,
- by determination of such sum as is reasonable.

MF/3 and G/90 being based on MF 'A' now look somewhat old fashioned in not providing for quotations but otherwise they are similar to MF/1 and MF/2.

Rates based or cost based valuations

It has been traditional for the valuation of variations to be based on rates or prices quoted in the contract. Then, depending upon whether the contractor's rates and prices are good or bad the contractor may make a profit or a loss in undertaking the variations. That is the nature of the bargain which has been struck with the purchaser.

There is a modern line of thought which has been taken up by the New Engineering Contract which says that a better method is to value all variations on a cost-plus basis. That effectively takes the financial risk out of variations and the purchaser pays the true price for his changes.

The model forms for plant contracts at first sight seem to operate on rates based rather than cost-plus valuations. The forms all refer to rates and prices. However if the schedule of prices is no more than a list of daywork rates the resultant valuation will be not far removed in its calculations from cost-plus.

A reasonable sum

The final fall-back position for valuing variations is expressed in most standard forms by reference to a reasonable sum. In the model forms for plant contracts this is put as a sum reasonable in all the circumstances.

The question is how is such a sum to be determined? If a contract price is high (or low) relative to a normal price is that contract's level of profitability (or loss) to be reflected in the valuation of variations? Is the value to the purchaser of a variation to be taken into account in what is reasonable? Should daywork rates or cost-plus be used to calculate the reasonable sum?

If there is any consensus of opinion on these matters it is probably that when a variation has to be valued on a reasonable sum basis – the preceding rules in the contract for valuation having proved inapplicable – then the contractor should at least receive his costs (including overheads) plus an allowance for profit.

From the contractor's viewpoint anything less would seem to be patently unreasonable but anything more difficult to justify.

However the recent case of *Laserbore Ltd* v. *Morrison Biggs Wall Ltd* (1993) involving microtunnelling works may now encourage contractors to be a little more ambitious in their expectation of what is a reasonable sum.

In the *Laserbore* case the work was carried out under the terms of a letter which stated that Laserbore would receive fair and reasonable payment for all work executed. Laserbore claimed on a rates basis using the FCEC (Federation of Civil Engineering Contractors) Schedules. Morrison Biggs Wall argued that payment should be on a cost-plus basis. The judge came down in favour of a rates based valuation and he rejected the cost-plus approach with this analysis:

'I am in no doubt that the costs plus basis in the form in which it was applied by the defendants' quantum experts (though perhaps not in other forms) is wrong in principle even though in some instances it may produce the right result. One can test it by examples. If a company's directors are sufficiently canny to buy materials for stock at knockdown prices from a liquidator, must they pass on the benefit of their canniness to their customers? If a contractor provides two cranes of equal capacity and equal efficiency to do an equal amount of work, should one be charged at a lower rate than the other because one crane is only one year old but the other is three years old? If an expensive item of equipment has been depreciated to nothing in the company's accounts but by careful maintenance the company continues to use it, must the equipment be provided free of charge apart from running expenses (fuel and labour)? On the defendants' argument, the answer to those questions is, "Yes". I cannot accept that that begins to be right.'

Costs of preparing quotations

In some contracts, including the IChemE model forms the contractor is given an express right to recover the costs of preparing quotations for variations on the instructions of the engineer whether or not, on consideration of the quotations, the variations are eventually ordered.

The model forms for plant contracts do not cover this point explicitly although MF/1 and MF/2 do both expressly allow the contractor to recover the costs of producing revised programmes.

For quotations for variations where the contract makes no express provision for payment the contractor's best case, if he wishes to seek recovery of the costs of preparation, is perhaps to argue that the engineer's instructions to provide quotations are themselves variations.

There is a sound case in this because although the model forms for plant contracts require the contractor to comply with instructions given by the engineer they do not directly address the question of payment for such instructions – which seems to imply that if they involve costs and are not related to the contractor's defaults they are variations.

16.5 Delay and disruption effects of variations

How should the delay and disruption effects of variations be valued?

Problem areas

Has the contractor got a contractual claim other than in the variation clause or can damages be sought for breach of contract? To what extent can the contractor roll together the delay and disruption effects of variations and present them as a global claim? Can the contractor claim that variations have so distorted the contract that the original price no longer applies and the whole of the work should be valued on a cost-plus or *quantum meruit* basis?

Basic principles

The starting points for the answers to these questions are quite simple if there is a variation clause in the contract with rules for valuation. Generally it will follow that:

- the contractor has undertaken to perform variations;
- the ordering of variations is not, of itself therefore, a breach of contract;
- claims relying on breach of contract are unlikely to succeed, consequently;
- where the contract provides the scheme for the valuation of variations there is no place for the introduction of extra-contractual schemes;
- rolled-up (or global) claims made outside the provisions of the contractual scheme for valuing variations usually suffer the difficulty of having no obvious contractual or legal basis;
- only in exceptional circumstances where the scheme for valuing variations on an individual basis has broken down can rolled-up (or global) claims be made within the provisions of the contract;
- claims that the contract has been distorted (or even frustrated) by the ordering of variations so as to justify abandonment of the contract price in favour of cost-plus or *quantum meruit* are unlikely to succeed. This is particularly so in the case of plant and process contracts which have limitations on the value of variations which may be ordered without the contractor's consent.

The McAlpine Humberoak case

Many of the basic principles listed above were examined in the case of *McAlpine Humberoak* v. *McDermott International* (1992). In that case the contract for the deck structure of an off-shore drilling rig was let on a lump sum basis. The contractor claimed extra costs arising from variations. The judge, at first instance, held that the contract had been frustrated by the extent of the variations and awarded the contractor *quantum meruit*. The Court of Appeal overturned this decision and held:

- the variations did not transform the contract or distort its substance or identity;
- the contract provided for variations;
- the contractual machinery for valuing variations had not been displaced;
- an award of *quantum meruit* could not be supported.

Time for ordering variations

Contractors sometimes allege that it is the time at which variations are ordered which leads to delay and disruption. From this they try to develop claims for breach of contract.

The legal rule would seem to be, however, that providing variations are ordered at a time which is reasonable in the circumstances there is no breach of contract.

Thus in *Neodox Ltd* v. *Swinton & Pendlebury BC* (1958) the contractor alleged an implied term that instructions would be given to enable him to complete in an economic and expeditious manner.

It was held by Mr Justice Diplock that under the terms of the contract it was clear that instructions would be given from time to time and what was reasonable did not depend solely on the convenience and financial interests of the contractor. He said:

'To give business efficacy to the contract details and instructions necessary for the execution of the Works must be given by the Engineer from time to time in the course of the contract. If he fails to give such instructions within a reasonable time the Corporation are liable in damages for breach of contract.'

And in *A. McAlpine & Son* v. *Transvaal Provincial Administration* (1974), a South African case, a motorway contractor asked the court to define an implied term on the time for supplying information and giving instructions on variations as either:

● a time convenient and profitable to himself;
● a time not causing loss and expense; or
● a time so that the works could be executed efficiently and economically.

The court declined on the grounds that under the contract variations could be ordered at any time irrespective of the progress of the works and that drawings and instructions should be given within a reasonable time after the obligation arose.

16.6 *Variations in MF/1*

Clause 27.1 – Meaning of variation

In these Conditions the term 'variation' means any alteration of the Works whether by way of addition, modification or omission.

This definition needs to be read in conjunction with Clause 1.1.p which defines the Works:

'Works means all Plant to be provided and work to be done by the Contractor under the contract.'

Note that the defined term 'variation' is not restricted to variations ordered by the engineer. However the valuation of variation rules and the extension of time provisions do only apply to ordered variations. But in any event see the expected 1995 change to clause 27.2 mentioned below.

The definition of variation does not expressly include, as in some contracts, matters as timing and methods of construction. For such matters which are not dealt with elsewhere in MF/1 (and some of course are dealt with as suspensions or mitigation of delay) it may be possible to imply that they are variations.

Clause 27.2 – engineer's power to vary

The Engineer shall have the power by notice to the Contractor from time to time during the execution of the Contract to instruct the Contractor to make any variations to the Works.

The Contractor shall carry out such variations and be bound by these Conditions in so doing as though the variations were stated in the Specification.

As soon as possible after having received any such instruction, the Contractor shall notify the Engineer if, in the Contractor's opinion, the variation will involve an addition to or deduction from the Contract Price.

No such variation shall together with any variations already ordered, involve a net addition to or deduction from the Contract Price of more than 15 per cent thereof unless the Contractor and the Purchaser consent thereto in writing.

Nothing in this Sub-Clause shall prevent the Contractor from making proposals to the Engineer for variations to the Works, but no variation so proposed shall be carried out by the Contractor except as directed in writing by the Engineer.

'The Engineer shall have power' indicates the exercise of discretion but as discussed earlier in this chapter there may be situations where the engineer is under an obligation to order variations.

The words 'power by notice' need to read in conjunction with clause 1.4 and 10.2. Clause 1.4 requires notices to be in writing unless otherwise specified and clause 10.2 requires notices to be sent to a nominated address.

'During the execution of the Contract' suggests that variations can be ordered during the defects liability period. Had the word 'Works' been used instead of 'Contract' the matter might be different.

It is unlikely that the requirement for the contractor to give notice if the variation will involve addition to or deduction from the contract price is a strict condition precedent to a change in the price. If it was, the contractor could avoid deductions from the contract price for omission variations simply by failing to give notice.

It is understood that the 1995 Revision 3 to MF/1 will make a significant change to the opening paragraph of clause 27.2 by adding words to the effect that the engineer alone shall have power to order variations until the works have been taken-over. This will sensibly qualify the presently over-wide meaning of variation in clause 27.1.

Financial limits

The intention of the 15% net limit on the value of variations is probably that the contract price as originally stated shall not rise or fall by more than 15%. But this interpretation would allow an omission variation of 30% if there had previously been addition variations amounting to 15%. It is therefore possible to argue an alternative interpretation of the words 'a net addition to or deduction from' such that the total of additions must not exceed 15% and the total of deductions must not exceed 15%.

It is not clear if the consents of the contractor and the purchaser required for variations exceeding the 15% limit are subject to the provision in clause 1.4 that any consent shall not be unreasonably withheld. Nor is it clear to whom the consents of the contractor and the purchaser are to be given.

As to how the consents of the parties on relaxation of the 15% limits are obtained is also unclear. Neither party is under any contractual obligation to raise the matter but both parties are assumed to have knowledge of the value of a variation before it becomes effective.

In the event of a variation proceeding without the necessary consents because one or both of the parties has failed to recognize that the limits will be exceeded it is possible that on the strict wording of clause 27.2 the variation is invalidated and the extra work falls to be valued outside the terms of the contract. However, the wording of clause 27.6 seems to indicate that all variations are valid unless the contractor raises the matter of the 15% limits with the engineer. On balance it is suggested that clause 27.6 provides the better test.

The question of whether contractor proposed variations come within the scope of the 15% limits may turn on the use of the word 'directed' in respect of such variations in contracts with the word 'instruct' for other variations. That is to say, variations which have been instructed clearly come within the limits but variations which are directed may not.

Clause 27.3 – Valuation of variations

> The amount to be added to or deducted from the Contract Price shall, if not the subject of a quotation from the Contractor which has been accepted by the Purchaser prior to the variations having been ordered, be determined in

accordance with the rates specified in the schedules of prices, if applicable. Where rates are not contained in the said schedules or are not applicable then the amount shall be such sum as is in all the circumstances reasonable. Due account shall be taken of any partial execution of the Works which is rendered useless by any such variation.

Note that acceptance of a quotation is by the purchaser and not by the engineer. As the clause is worded above there is even some doubt as to the role of the engineer in determining values not fixed by quotations. The engineer is not given any specific power to value variations and the intention of the contract would seem to be that the contractor and purchaser strive to reach agreement failing which they go to arbitration.

However, it is understood that the 1995 Revision 3 to MF/1 will clarify the point that value of variations is to be determined by the engineer.

Clause 27.4 – contractor's records of costs

In any case where the Contractor is instructed to proceed with a variation prior to the determination of the value thereof under Sub-Clause 27.3 (Valuation of Variations) the Contractor shall keep contemporary records of the Cost of making the variation and of time expended thereon. Such records shall be open to inspection by the Engineer at all times.

This clause seems to imply that where the value of a variation is not fixed before the instruction to proceed is given the variation is to be valued on a cost basis. This, however, is not compatible with the wording of clause 27.3 which requires rates and prices to be used if applicable. The explanation is presumably that cost records should be kept in case the rates and prices are not applicable and a reasonable sum becomes due.

It is questionable whether the clause 27.4 requirement of keeping contemporary records and making them available for inspection extends to sub-contractors and suppliers – although clause 11.1 of the model form of sub-contract, if used, does oblige sub-contractors to do so.

Clause 27.5 – notice and confirmation of variations

When ordering any variation to any part of the Works, the Engineer shall give the Contractor such reasonable notice as will enable him to make his arrangements accordingly.

In cases where Plant is already manufactured, or in the course of manufacture, or any work done or drawings or patterns made that require to be altered, the Contractor shall be entitled to be paid the Cost of such alterations.

If, in the opinion of the Contractor, any such variation is likely to prevent or prejudice him from or in fulfilling any of his obligations under the Contract, he shall notify the Engineer thereof with full supporting details. The Engineer shall decide forthwith whether or not the variation shall be carried out.

If the Engineer confirms his instructions in writing the said obligations shall be modified to such an extent as may be justified. Until the Engineer so confirms his instructions, they shall be deemed not to have been given.

The contractor's remedy if the engineer fails to give reasonable notice would appear to be no more than to include any costs arising from such failure in the costs of the variation. As explained elsewhere in this book the contractor cannot found a claim on breach of contract under MF/1.

The contractor's entitlement to be paid 'the Cost' of alterations does not fit readily with the rules for valuation of variations in clause 27.3. However, the words of clause 27.5 are clear enough and there seems little doubt that alteration work is paid for on a cost basis.

The words 'any such variation' used in relation to fulfilment of the contractor's obligations may be misleading in that the intended application is to 'any variation'. The word 'such' suggests the provision applies only to alterations.

The obligations referred to in the clause can be identified from clause 13.1 (contractor's general obligations). These cover design, manufacture, delivery, erection, completion on time and other things.

The requirement for the contractor to give notice if a variation is likely to prevent or prejudice any of these obligations can be seen as protecting the interests of both the contractor and the purchaser. Although the purchaser may have no positive remedy against the contractor for failing to give notice, the contractor might well find that such failure could adversely affect his entitlements or alter his obligations under the contract. For example, in respect of an extension of time for completion or in relation to performance.

For the engineer to decide 'forthwith' whether or not any variation should be carried out may be no easy matter if technical objections have been raised by the contractor. 'Forthwith' in this context can mean no more than getting on with the matter promptly.

The intention of the phrase 'the said obligations shall be modified' is clearly to preserve the position of the contractor if variations are ordered notwithstanding the contractor's expressed concerns on their effects on his obligations. That is an entirely reasonable approach. However what the clause does not do is clarify whether the obligations are to be formally modified or whether they are implied to have been modified should a dispute on obligations subsequently arise. Formal modification, if it can be agreed, is obviously preferable but implied modification leaves the contractor with more scope for argument.

The final provision of clause 27.5 that instructions are deemed not to have been given until confirmed by the engineer has the effect of leaving a contractor who has questioned the effects of variations on his obligations without any instructions until they are reconfirmed. It is not open to a contractor, therefore, to give the engineer his opinion of the effects of an ordered variation and then to proceed on the basis that the engineer has been warned of the outcome. The contractor in such a position is proceeding with a change which is not an authorized variation.

Clause 27.6 – progress with variations

> The Contractor shall on receipt of the Engineer's instructions under Sub-Clause 27.2 (Engineer's Power to Vary), or confirmation of instructions under Sub-Clause 27.5 (Notice and Confirmation of Variations), immediately proceed to carry out such instructions, unless the Contractor has notified the Engineer that the variation in his opinion will involve a net addition to or deduction from the Contract Price of more than 15 per cent.
>
> Subject thereto, the carrying out of such instructions shall not, without the consent of the Engineer, be delayed pending agreement on price

Clause 27.6 does not indicate what follows notification by the contractor that a variation will, in his opinion, exceed the 15% limits.

As mentioned above the contractor is not under any express obligation to give notice of a potential breach of the limits but both parties are assumed to have enough knowledge of the value of a variation to give their consents should they wish the variation to proceed. However, only the contractor is entitled under clause 27.6 to delay the progress of a variation. It is not clear why the purchaser has no say in the matter – unless, perhaps, his consent is deemed to have been given.

If agreement is not reached between the parties so that either both parties' consents are given or the value of the variation is settled within the 15% limits there is nothing in the contract short of arbitration to resolve the matter. In practical and timing terms this may be a very poor solution.

16.7 Variations in MF/2

The text of clause 21 of MF/2 is the same as the text of clause 27 of MF/1 and consequently it is not reproduced in this section.

The obligations of the contractor under MF/2 are, of course, less than those under MF/1 and this correspondingly reduces the scope of the variation clause. But this apart, there is nothing in the operation of clause 21 of MF/2 to distinguish it from the operation of clause 27 of MF/1.

16.8 *Variations in MF/3*

Clause 13.1 – variations

> The Vendor shall not alter any of the goods, except as directed in writing by the Purchaser; but the Purchaser shall have full power, subject to the proviso hereinafter contained, from time to time during the execution of the Contract by notice in writing, to direct the Vendor to alter, amend, omit, add to or otherwise vary any of the goods and the Vendor shall carry out such variations, and be bound by the same conditions, so far as applicable, as though the said variations were stated in the Specification. No such variation shall together with any variations already ordered, involve a net addition to or deduction from the Contract Price of more than 15% unless Vendor and Purchaser so agree in writing.

Variation is not defined as in MF/1 and MF/2 but the words 'alter, amend, omit, add to or otherwise vary' are just as wide.

The requirement for the Vendor and Purchaser to agree before ordering variations which exceed the 15% valuation limit is more positive than the arrangement in MF/1 and MF/2 for consents.

Clause 13.2 – valuation of variations

> In any case in which the Vendor has received any such direction from the Purchaser which either then or later will, in the opinion of the Vendor, involve an addition to or deduction from the Contract Price, the Vendor shall, as soon as reasonably possible, advise the Purchaser in writing to that effect. The amount to be added to or deducted from the Contract Price shall be ascertained and determined in accordance with the rates specified in the schedules of prices, so far as the same may be applicable, and where rates are not contained in the said schedules, or are not applicable, such amount shall be such sum as is reasonable in the circumstances. Due account shall be taken of any partial execution of the Contract which is rendered useless by any such variation.

There is a significant difference here from MF/1 and MF/2 in that there is no express reference to valuation by quotation. Otherwise the rules on valuation are the same.

Clause 13.3 – notice and confirmation of variations

> If the Purchaser shall direct that any such variation be made, such reasonable notice in writing shall be given to the Vendor as will enable him to make his arrangements accordingly. If in the opinion of the Vendor any such variation is likely to prevent or prejudice the Vendor from or in fulfilling any of his obligations under the Contract, he shall notify the Purchaser thereof in writing, and

the Purchaser shall decide forthwith whether or not the same shall be carried out. If the Purchaser confirms his instructions in writing, the said obligation shall be modified to such an extent as may be justified. Until the Purchaser so confirms his instructions they shall be deemed not to have been given.

The provisions here correspond very closely with the provisions in MF/1 and MF/2.

16.9 Variations in G/90

Clause 10(i) – variations and omissions

The Contractor shall not alter any of the Works, except as directed in writing by the Engineer; but the Engineer shall have full power, subject to the proviso hereinafter contained, from time to time during the execution of the Contract by notice in writing to direct the Contractor to alter, amend, omit, add to, or otherwise vary any of the Works (including the sequence and timing thereof), and the Contractor shall carry out such variations, and be bound by the same conditions, so far as applicable, as though the said variations were stated in the Specification. Provided that no such variation shall, except with the consent in writing of the Contractor, be such as will, with any variations already directed to be made, involve a net addition to or deduction from the Contract Price of more than 15 per cent thereof. If any case in which the Contractor has received any such direction from the Engineer which either then or later will, in the opinion of the Contractor, involve an addition to or deduction from the Contract Price, the Contractor shall, as soon as reasonably possible, advise the Engineer in writing to that effect. The amount to be added to or deducted from the Contract Price shall be ascertained and determined in accordance with the rates specified in the schedules of prices, so far as the same may be applicable, and where rates are not contained in the said schedules, or are not applicable, such amount shall be such sum as is reasonable in the circumstances. Due account shall be taken of any partial execution of the Works which is rendered useless by any such variation.

If the Engineer shall make any such variation in any part of the Works, such reasonable notice in writing shall be given to the Contractor as will enable him to make his arrangements accordingly. If in the opinion of the Contractor any such variation is likely to prevent or prejudice the Contractor from or in fulfilling any of his obligations under the Contract, he shall notify the Engineer thereof in writing, and the Engineer shall decide forthwith whether or not the same shall be carried out. If the Engineer confirms his instructions in writing, the said obligations shall be modified to such an extent as may be justified. Until the Engineer so confirms his instructions they shall be deemed not to have been given.

This is essentially the same clause as in MF/3 – both are taken from the old MF 'A'. Much of the detailed comment in Section 16.6 on MF/1 applies to these provisions.

Chapter 17

Claims for additional payments

17.1 Introduction

Unlike the many standard forms of contract used in the construction industry the model forms for plant contracts have traditionally had little to say on the subject of claims for additional payments. The various editions of MF 'A' never went far beyond identifying a limited number of situations which could attract payment in like manner to variations and this is the approach still followed in G/90.

MF/1 however departs from this policy. It sets out in clause 41 the procedures and notice requirements for claims and it provides for recovery on a cost basis – with allowance for profit permitted in some circumstances. Cost is made a defined term with specific reference to overheads and financing charges.

MF/2 follows the policy of MF/1 but covering as it does only delivery of plant it gives fewer express entitlements to additional payments. MF/3, covering the delivery of goods, does not contain a detailed claims procedure although it does allow for some changes in the contract price.

Grounds for additional payment

In both MF/1 and G/90 the express grounds for additional payments fall into the following broad categories:

- variations
- unexpected site conditions
- interference with tests
- delayed delivery and/or suspension
- incorrect information
- breach of the purchaser's obligations
- changes in statutory regulations
- adjustments for provisional sums and prime cost items
- price fluctuations
- other specified entitlements.

MF/1 is more detailed in its provisions than G/90 but the model forms take a similar approach to the allocation of risk between the parties and the purchaser generally takes much the same responsibility for his own defaults and those of the engineer under both forms. However, when it comes to entitlements to additional payments which are not expressly covered in the contracts there is a significant difference between MF/1 and G/90.

This arises because one effect of clause 44.4 of MF/1 (exclusive remedies) is to severely limit, if not altogether to exclude, the contractor's common law remedies for breach of contract by the purchaser. That is to say, under MF/1 the contractor has the express rights as set out in the contract but no other rights. So, arguably, the contractor cannot claim for breach of an express term of the contract unless payment is provided for in the contract nor can he claim for breach of an implied term of the contract.

This is not a common contractual situation and because of its importance clause 44.4 is discussed in some detail later in this chapter and elsewhere in this book.

Lump sum pricing

Plant contracts are usually priced on a lump sum basis and MF/1, MF/2, MF/3 and G/90 are all intended as lump sum contracts.

For the purchaser, such contracts have the attraction of greater certainty of price than remeasurement or cost reimbursable contracts and for that reason alone the appeal of lump sum pricing is obvious. There is, however, a further advantage to the purchaser in entering into lump sum contracts in that they avoid the numerous grounds for additional payment which can be generated out of remeasurement contracts with schedules of rates or bills of quantities.

Users of the Additional Special Conditions Supplement (S1) to MF/1 which deals with payments in whole or part by measurement need to be aware of the dangers of claims arising from loosely specified methods of measurement or inadequately prepared schedules or bills.

Responsibility for design

In many contracts claims for additional payments arise out of the purchaser's or the engineer's responsibility for design. Whether it be in the timing of the supply of design information to the contractor or in the adequacy of that information the potential exists for the contractor to allege that in consequence of some default he has suffered additional costs and that recovery of such cost should be available either under the provisions of the contract or as a claim for breach of contract.

Clearly, the less responsibility the purchaser and his engineer take for design, and the more is given to the contractor, the less the contractor's opportunities for claims.

In plant contracts, the contractor usually takes all, or most, of the responsibility for design and accordingly express provisions for claims relating to design are usually minimal or missing. G/90 does give the contractor entitlement to costs caused by errors or omissions in information supplied by the purchaser or the engineer but it says nothing about late information. MF/1 is similar to G/90 but there are also conflicting provisions in the form on the contractor's responsibility for design supplied by the purchaser or the engineer. These appear to make a distinction between design supplied to the contractor and information supplied to the contractor. More is said on this later in this chapter.

The amount recoverable

Claims for additional payment may, according to what the contract prescribes for a particular situation, be either cost based or rates based.

Some contracts such as the New Engineering Contract rely entirely on cost-based recovery but more commonly the valuation of variations is rates based and payment for unforeseen conditions or interference claims is cost based. MF/1 generally follows this latter approach but G/90 however follows the scheme in the old MF 'A' and links most entitlements to recovery to the rules for the valuation of variations.

In practice, the difference is not always significant because unless there are applicable rates within the contract the contractor is entitled to a reasonable sum – which is likely to be cost based. A point of more significance is what is meant by, or what is due under the definition of cost?

MF/1 provides a definition of cost which includes all expenses and costs including overhead and financing charges whilst excluding profit. But even this leaves open to argument what overheads are recoverable and how they are to be verified. G/90 does not define cost although the phrase in clause 10(1) 'such sum as is reasonable in the circumstances' is an adequate benchmark for most claims.

The contractor may, of course, argue that a sum which is reasonable should include an allowance for profit – and in the context of clause 10(1) of G/90 (the variation clause) that is not an unreasonable proposition. As a general rule, however, cost and profit are distinguishable. And although profit, or loss of profit, is recoverable in some breach of contract claims the position under most contractual entitlements is that an allowance for profit is only claimable where there is an express provision for profit.

MF/1 addresses this issue particularly well and states in clause 41.2 precisely when profit is allowable on claims. G/90 does not mention profit.

Delay and disruption

The question of how and when a contractor can recover the costs of delay and disruption is left open in most standard forms of contract. Many forms avoid the issue altogether and there is certainly very little by way of specific guidance in either MF/1 or G/90.

In MF/1 the costs of delay and disruption have to be taken in with the costs of claims made under the express provisions of the contract – because all other claims are excluded. In G/90 the position is more flexible and delay and disruption claims of a general nature could succeed in appropriate circumstances.

Notification of claims

As a general rule, claims for breach of contract are not governed by notice requirements. Providing action on such claims is brought within the statutory limitation periods the claims stand or fall on their merits.

Contractual entitlements to additional payments can however, and frequently do, contain notice requirements which are binding on the contractor. A claim which might be valid in substance may then be disallowed as out of time.

MF/1 has perhaps the strictest notice regime of any standard form of contract and the purchaser has no liability for claims unless notice is given within 30 days of the circumstances arising. G/90 by contrast has no notice requirements but this is probably more because there is no specific claims clause than by way of intentional indulgence.

Valuation and settlement of claims

It is a matter of some importance to contractors, and no doubt to a lesser extent to purchasers, how the value of claims is to be assessed and how disputes on an assessment are to be settled. In many recently issued standard forms of contract elaborate procedures have been devised for both valuation and dispute resolution usually with the intention of securing early assessment of the amounts due and avoiding arbitration if at all possible.

MF/1 retains the traditional approach of valuation by the engineer subject to review in arbitration if the parties are in dispute. However on this it should be said that the power of the engineer to assess claims is limited to claims made under the provisions of the contract – although with MF/1, since no other claims are permitted, this is not the problem it is under other standard forms.

G/90 does not state who is to assess claims or who is to value variations. The various clauses with provisions for additional payments simply refer to amounts being ascertained. Even the final certificate which is stated in clause 31(ix) to be conclusive evidence as to the value of the Works appears to rely on the application of the contractor rather than on the opinion of the engineer as to what is due.

No doubt it can be argued that the power of the engineer to value claims and variations is implied in G/90 but an alternative argument is that the purchaser must accept the contractor's applications for additional payments unless and until they are amended by an arbitrator.

17.2 *Legal basis of claims*

Claims for additional payments will either be claims under the terms of the contract or claims at common law. These are sometimes known as contractual claims and extra-contractual claims.

Contractual claims

The legal basis of contractual claims is the contract itself. The claiming party, usually the contractor, simply has to establish that the express conditions for additional payment have occurred in order to acquire an entitlement to payment.

Subject to considerations of legality, the parties are free to contract on whatever terms they wish. They cannot contract out of statutory obligations or restrictions but they can, and frequently do, depart from the rules of common law in apportioning risk and responsibilities.

So it will often be the case that contractual claims can be made where there would be no common law entitlement. For example in MF/1 and G/90 the contractor can claim for the additional costs of unforeseen ground conditions and for additional costs arising from changes in statutory regulations – neither of which could be the basis of a common law claim for breach.

Contrary to a quite popular belief contracts need not be fair. The Unfair Contracts Terms Act 1977 sometimes mentioned in this context has more to do with preventing a party from excusing itself from its own negligence than with any general concept of fairness. But again contrary to an even more popular belief amongst contractors most plant, process and construction contracts are fair in the sense that they give contractors more opportunities to claim additional payments than does the common law.

Contractual claims also have other advantages for contractors in that interim payments may be permitted and the basis of recovery may be clearly defined. Against this, and leaning towards the advantage of the purchaser, is that contractual entitlements to payment may have strict notice and other administrative requirements.

Common law claims

In the main, common law claims depend upon establishing a breach of contract.

The breach will usually be related to an express term of the contract which places on one party an obligation to the other. For example, to give possession of the site or to obtain planning consent.

But the breach may also be related to a term implied into the contract to give it business efficacy. For example, that one party will not hinder the other in the performance of the contract. However it is not the purpose of implied terms to improve the contract the parties have made and a term will not be implied when there are express terms in the contract dealing with the point at issue.

Contracts rarely provide any provisions for the submission, valuation or payment of common law claims and strictly speaking they are claims which should be dealt with outside the contract directly between the contractor and the purchaser – hence the name, extra-contractual claims.

One aspect of these claims, which is frequently overlooked, is that the engineer is most unlikely to have any power or authority under the contract to deal with them. The powers and authority of the engineer derive solely from the contract and, as a general rule, they do not extend to extra-contractual matters.

In practice, however, engineers do often become involved in claims made on a common law basis. This is partly because it is not always clear on what basis a claim is being made. It is also partly because the engineer is often the administrative link between the contractor and the purchaser.

Where the engineer does become so involved he is in effect acting as the agent of the purchaser. In this capacity he is not required to be impartial or to make a fair valuation. His task will be to act in the best interests of the purchaser. This probably involves getting the best settlement he can for the purchaser.

Alternative claims

Most standard forms of contract, including MF/1 and G/90 give contractual rights to additional payments for some circumstances which could be classed as clear breach of contract. Failure to give possession of the site is the most obvious example.

The question which then arises is does the contractor have a choice? Must he seek additional payment under the terms of the contract or can he elect to sue for breach of contract and recover his damages as a common law claim? The question is relevant in relation to giving, or failing to give, notice of intention to claim and in relation to the amount recoverable where the contractual amount is less than the amount recoverable as damages.

In MF/1 and MF/2 for example the question is of particular significance. In both these forms the purchaser is not liable to make payment in respect of any claim for additional payment made under the contract unless the contractor has strictly complied with specified notice requirements. Consequently unless there is an alternative common law right of claim the contractor has no remedy if he fails to give the specified notice.

The general rule is that common law rights are only excluded if there are express terms in a contract to that effect. Few standard forms do this in a positive way and some, such as the building form JCT 1980, expressly state the opposite – that the provisions of the contract are without prejudice to other rights and remedies.

G/90 is silent on the subject and the common law rights of the contractor to claim for breach of contract probably remain effective. The only area of doubt is whether a contractual entitlement is of itself to be regarded as an express exclusion of the alternative common law right. The point is arguable but it lacks firm legal backing.

MF/1 and MF/2 however are amongst the few standard forms which do expressly exclude common law rights. Clause 44.4 of MF/1 and clause 32.4 of MF/2 both make it clear that not only are the rights of the parties as provided in the contracts the exhaustive rights but also that neither party is liable to the other for any breach of contract except as expressly provided for in the contract.

In other words, in MF/1 and MF/2, there can be no alternative common law claims where there are contractual entitlements to extra payments nor can there be common law claims where there are not contractual entitlements. The contracts, by themselves, provide the exclusive remedies.

Implied terms

One effect of the 'exclusive remedy' provisions of MF/1 and MF/2 is that claims cannot be founded in these contracts on implied terms. This can operate to the serious disadvantage of the contractor.

Very few contracts express in full the obligations of the parties or state for each obligation that is expressed that a contractual remedy exists in the event of non-fulfilment. Even to attempt such a task would normally be seen as over ambitious and wholly unnecessary. This is because terms will be implied into contracts to give them business efficacy.

For example, in G/90 under clause 6(ii) the contractor is not permitted to sub-let without the consent of the engineer which shall not be unreasonably withheld. The clause does not expressly state that the engineer shall consider applications to sub-let which are put to him by the contractor. But it goes without saying that the engineer has this duty. This is implied; as is a term that should the engineer neglect to perform his duty and thereby obstruct the contractor then the purchaser would be liable in damages.

Under MF/1 similar circumstances can be envisaged – clause 3.2 prohibits sub-contracting without the prior consent of the engineer. However even though clause 2.1 states that the engineer shall carry out the duties specified in the contract, the contractor is not given any express entitlement to additional payment for default by the engineer. An implied term would overcome this. But, as stated above, without being able to rely on an implied term the contractor is seriously disadvantaged.

17.3 The amount recoverable

The amount due to the contractor by way of additional payment will be calculated either under the provisions of the contract or as damages for breach of contract depending upon the legal basis of the claim. In exceptional circumstances, discussed briefly later in this chapter, payment outside the contract on a *quantum meruit* basis may become due.

Payment under the contract

Calculations made under the contract must be in accordance with the wording of the contract. This is not always as straightforward as it seems because in most contracts the phrasing of the entitlement varies from clause to clause and even where there are defined terms such as 'cost' there is often scope for argument as to what can be properly recovered.

On the phrasing point, the model forms for plant contracts show more discipline than many construction contracts but nevertheless there is still more variety than appears to be strictly necessary.

In MF/1 the following phrases are used:

- Clause 5.7 (unexpected site conditions) – extra costs.
- Clause 6.1 (statutory or other regulations) – the amount of any increase or reduction.
- Clause 11.8 (breach of the Purchaser's general obligations) – additional cost reasonably incurred.
- Clause 25.2 (suspension) – any additional cost incurred.

The change in wording from clause 5.7 to clause 6.1 is understandable,

since clause 6.1 is effectively an extension of the price fluctuations provision to cover all financial risk arising from statutory changes occurring
after the date of the tender. However, it is questionable if a distinction is
necessary or is even intended between the phrases used in clauses 5.7,
11.8 and 25.2.

It can certainly be argued that there is a difference between 'cost
reasonably incurred' and 'any additional cost'. The former would seem to
bring in concepts of remoteness and mitigation whereas at face value the
latter gives an unqualified entitlement.

In G/90 the majority of the additional payment provisions refer
somewhat unusually (but following the wording of the old MF 'A') to 'a
sum ascertained and valued in like manner to the valuation of variations'.

This does suggest that additional payment claims, including some
dealing with what amounts to purchaser's breach of obligations, are to be
valued on a rates basis instead of a cost basis. Indeed, if the contract does
contain a schedule of rates for the valuation of variations – which may be
no more than a daywork schedule – then that should be used.

However, in many cases the schedule of rates will be inapplicable and
in this case G/90 provides (in clause 10(i)) that the contractor's entitlement is 'such sum as is reasonable in the circumstances'. This will more
often than not bring the valuation back to a cost basis.

The meaning of cost

In MF/1 and MF/2 cost has a defined meaning. In both model forms
clause 1.1.j reads:

'Cost means all expenses and costs incurred including overhead and
financing charges properly allocable thereto with no allowance for
profit.'

G/90 has no definition of cost but this is not too significant because the
form only uses the word sparingly. Extra cost is given as the basis of
additional payment in clause 5(ii) (mistakes in information supplied by
the purchaser) but elsewhere, as mentioned above, most additional
payments are to be valued as variations.

The meaning of cost it is suggested, whether defined or undefined, is
different from the meaning of loss and expense – the corresponding
phrase used in the building industry. The difference is not always
recognized and contractors under plant and process contracts frequently
slip into the use of loss and expense terminology when making claims.
But whereas the meaning of loss and expense has been examined by the
courts and held, in the case of *Wraight Ltd* v. *P.H. & T. (Holdings) Ltd* (1968)
to be equivalent to damages recoverable at common law there is no

similar legal ruling relating to cost to suggest that the meaning of cost is wide enough to include for loss.

The point is important because many contractors' claims really are for loss rather than for cost. For example, claims for head office overheads for prolongation as presented using Hudson's or Emden's formulae are loss of opportunity claims not cost claims.

In MF/1 the inclusion of the word 'incurred' in the definition of cost adds weight to the argument that the contractor's entitlement to recovery does not extend to loss.

Although similar wording is found in other contracts, most notably the ICE Conditions, the effect is different in MF/1 from in most other contracts. This is because, as mentioned earlier, MF/1 excludes common law claims for breach – which can include for loss – whereas most other contracts leave the contractor with common law remedies.

Recovery of overheads

Overheads are normally categorized as being on-site or off-site, the latter being another name for head office overheads.

Claims for additional cost of on-site overheads are straightforward providing that adequate records exist to identify the items and amounts in the claim. There is however sometimes evidence of items being claimed both as on-site and off-site overheads. And since there is not common classification of items between these headings, engineers need to be alert to the possibility of duplication.

As a generalization, large contractors put more items into on-site overheads than do small contractors – aided, no doubt, by more sophisticated accounting systems. In consequence, large contractors usually have a higher level of on-site and correspondingly lower level of off-site overheads than small contractors.

Claims for off-site overheads can be far more complex than claims for on-site overheads. There is no objection in principle to such claims whether they be claims under the contract – with overheads included in the definition of cost as in MF/1 – or they be common law claims. The difficulty is in evaluating the extra cost incurred as a result of the claimable event.

Thus in the case of *Tate & Lyle Ltd* v. *Greater London Council* (1982) it was held that in principle Tate & Lyle could recover damages for managerial and supervisory expenses but it was not permissible simply to add a percentage to other items in the claim. It was said that modern office arrangements should permit the recording of time spent by managerial staff on a particular project.

Put simply, the point is should head office overheads be treated any differently from other heads of claim in the detail of proof and records of extra cost required to justify additional payment? The *Tate & Lyle* case suggests that the answer is no; but applied strictly that would deprive contractors in most claims of any recovery for head office overheads.

The problem is that most contractors would be unable to prove either any extra head office overhead costs as a result of a claimable event or would be unable to show any allocation of managerial time to the event. Head office overhead claims for prolongation are, as mentioned above, different in that they are essentially loss claims and calculation of the amount due, in those circumstances where the claim is permissible, is likely to be on a theoretical basis.

Under MF/1 and MF/2 there is an argument that cost as defined in the contract permits overheads to be allocated instead of proved. The words used in clause 1.1.j are 'including overhead and financing charges properly allocable'. Perhaps if the matter ever came to be argued in court the interpretation would turn on the significance of the word 'properly'. This it might be said does indicate need of proof. Against this it can be said that it does not seem to be the practice in the industry to require such proof and allocation of head office overheads by percentage is normal.

However, even when head office overheads are accepted on a percentage basis there is still the matter of proof of the percentage claimed. The best method, it is suggested, is to extract figures from audited accounts. This at least shows some relationship with the contractor's actual costs – albeit on a companywide basis. The alternative method of using tender based percentages has the weakness that there is no tangible connection with actual cost.

Financing charges

The definition of cost in MF/1 and MF/2 includes for financing charges. Both model forms also allow for interest to be claimed on delayed payments and it might seem on the face of it that financing and interest charges are much the same thing. However in law they are different.

Interest charges relate to late payment of a sum due. Financing charges relate to costs which form the subject of claims for the period from which such costs are incurred to the time of application for payment or certification.

The principle concerned in the payment of financing charges is that the contractor has incurred cost in financing the primary cost involved in his claim and this financing charge, therefore, is not interest on a debt but a constituent part of the debt.

The principle was established by the Court of Appeal in the case of *F. C. Minter Ltd* v. *Welsh Health Technical Services Organisation* (1980). The decision in the *Minter* case was followed by the Court of Appeal in *Rees and Kirby Ltd* v. *Swansea City Council* (1985) where it was confirmed that calculation of financing charges should be on a compound interest basis.

It is now generally accepted, following the *Minter* and *Rees and Kirby* cases that, under most standard forms of construction contracts, the contractor is entitled to include in his application for extra cost the financing charges he has incurred up to that time and this would seem to apply also to plant and process contracts.

MF/1 and MF/2 put the matter beyond doubt by expressly providing for financing charges. But with G/90 which is silent on the issue there is still resistance in some quarters to recognizing that such charges are a legitimate part of cost.

Profit

The question of whether it is permissible for a contractor to include a profit margin in a claim for additional payment can only be answered by reference to the terms of the contract and the legal basis of the claim.

In MF/1 and MF/2 the definition of cost excludes any allowance for profit and consequently profit is only payable under these model forms when elsewhere in the contract – clause 41.2 in MF/1 and clause 29.2 in MF/2 – an allowance for profit is specifically allowed on extra cost.

In fact the provisions in MF/1 and MF/2 allowing for profit are quite generous compared with other standard forms of contract. Not only is profit allowed on a wide range of claims but profit can be added to all extra cost in those claims – not merely on the additional work element as is usually the case.

The claims under MF/1 on which profit is allowed are listed in clause 41.2 as:

- Sub-Clause 5.7 (Unexpected Site Conditions)
- Sub-Clause 11.8 (Breach of Purchaser's General Obligations)
- Sub-Clause 14.5 (Revision of Programme)
- Sub-Clause 16.2 (Errors in Drawings, etc Supplied by Purchaser or Engineer)
- Sub-Clause 21.2 (Special Loads)
- Sub-Clause 22.1 (Setting Out)
- Sub-Clause 25.2 (Additional Cost Caused by Suspension)
- Sub-Clause 25.5 (Resumption of Work, Delivery or Erection)
- Sub-Clause 27.5 (Notice and Confirmation of Variations)
- Sub-Clause 31.2 (Tests during Defects Liability Period)
- Sub-Clause 33.3 (Mitigation of Consequences of Delay)
- Sub-Clause 36.8 (Contractor to Search)
- Sub-Clause 40.3 (Remedies on Failure to Certify or Make Payment) or
- Sub-Clause 52.2 (Performance to Continue During Arbitration).

An apparent omission from the list is sub-clause 25.6 (effect of suspension on defects liability). This is included in the corresponding list in MF/2 and it is understood that it will be added to the MF/1 list in the 1995 Revision 3 to MF/1.

G/90 makes no express reference to profit. But since the basis of valuation of claims for extra payment is generally the same as that for the valuation of variations the contractor will in effect recover profit in his claims if the rates and prices used for the valuation are themselves profitable.

For common law claims, loss of profit resulting from a breach of contract is allowable as a head of damages. But it is not generally permissible to add profit to costs in a damages claim. The purpose of damages is to restore the financial position of the claimant not to improve it.

17.4 *Additional payments under MF/1*

In MF/1 the full range of express provisions relating to additional payments (and other adjustments of the contract price – including deductions) is as follows:

Table 17.1 Express provisions relating to additional payments under MF/1

Clause	Description	Entitlement
1.1.j	definition of cost	cost incurred including overheads and financing charges but not profit
5.4	provisional sums	not stated
5.5	prime cost items	net amount plus percentage in the appendix
5.7	unexpected site conditions	extra costs of dealing with obstructions or hazards
6.1	changes in statutory or other regulations	amount of any post-tender increase or reduction in the cost of performing the contract
6.2	changes in prices of labour, materials and transport	amount of any post-tender increase or decreases in prices but excluding increases due to the default or negligence of the contractor
11.8	breach of the purchaser's general obligations imposed by clause 11	the additional cost reasonably incurred by the contractor
14.5	revisions of programme	the costs of producing a revised programme required for reasons for which the contractor is not responsible

Table 17.1 Contd

Clause	Description	Entitlement
16.2	errors in drawings supplied by the purchaser or the engineer	the cost incurred in carrying out instructions on alterations or remedial works
18.4	opportunities for other contractors employed by the purchaser	such sums as may, in the opinion of the engineer, be reasonable in making available equipment or providing services as instructed
19.1	work done outside normal working hours by instruction	extra cost of the work so done unless necessary by the contractor's default to ensure completion on time
21.2	special loads	the cost of work certified by the engineer for strengthening highways or bridges (or the value if there are applicable rates)
21.3	extraordinary traffic claims	the purchaser indemnifies the contractor in respect of claims but the engineer may DEDUCT from the contract price amounts due to the contractor's negligence or failure to perform obligations
22.1	setting out errors resulting from incorrect information supplied by the purchaser or the engineer, or default by another contractor	the cost incurred in rectifying errors
23.5	failure on tests or inspection attributable to the contractor	all expenses incurred by the purchaser in retesting and inspection DEDUCTED from the contract price
25.2	instructions to suspend the progress of the works	any additional cost incurred by the contractor
25.5	resumption of work, delivery or erection	the costs of examining and making good
25.6	effect of suspension on defects liability	the additional cost of performing obligations for defects more than three years after the normal delivery date
26.1	defects before taking-over	the purchaser may make good defects if the contractor defaults at the cost of the contractor. But note no express provision to DEDUCT
27.3	valuation of variations	amount determined by: – quotation – applicable rates – such sum as is reasonable

Table 17.1 Contd

Clause	Description	Entitlement
27.5	alterations due to variations	the cost of alterations to plant, work or drawings
28.3	delayed tests	costs of delayed tests may be DEDUCTED from the contract price if due to the contractor's default
28.4	repeat tests	costs incurred by the purchaser in the repetition of tests may be DEDUCTED from the contract price
29.4	outstanding work	if the contractor fails to do outstanding work the purchaser may arrange for completion with the COST DEDUCTED from the contract price
31.2	interference causing tests on completion to be undertaken during defects liability period	costs incurred by the contractor
33.3	mitigation of the consequences of delay	costs incurred by the contractor in complying with the engineer's restrictions to overcome or minimize delay for which the contractor is entitled to an extension of time
34.1	liquidated damages for late completion	may be DEDUCTED from the contract price or paid by the contractor
34.2	prolonged delay	the LOSS suffered by the purchaser up to the amount stated in the appendix
35.8	failure to pass performance tests	within stipulated limits the contractor to PAY liquidated damages; outside stipulated limits the contract price to be REDUCED
36.5	delay in remedying defects	the costs incurred by the purchaser in remedying defects may be DEDUCTED from the contract price or paid by the contractor
36.8	searches for defects	the cost of work in searching for defects not the contractor's responsibility to be added to the contract price
40.2	delayed payment	interest of 2% above average base rates payable on delayed payments

Table 17.1 Contd

Clause	Description	Entitlement
40.3	failure to certify or make payment	the additional cost occasioned by stoppage and resumption consequent upon failure by the engineer to certify or the purchaser to pay
41.2	allowance for profit on claims	applies to 14 specified provisions for additional payment
43.3	damage to the works caused by the purchaser's risks	making good to be at the expense of the purchaser at a price agreed or failing agreement a sum as in all the circumstances is reasonable
46.4	payment on termination for *force majeure*	– the cost of goods ordered – other expenditure reasonably incurred – reasonable cost of removal of equipment – reasonable cost of repatriation of staff
48.1	failure to insure	the purchaser may effect insurances and DEDUCT the premiums from monies due to the contractor or recover as a debt
49.3	payment after termination due to the contractor's default	excess of the costs of completion over the amount payable to the contractor PAYABLE on demand to the purchaser
51.3	payment after termination due to the purchaser's default	termination value plus other expenditure reasonably incurred plus loss of anticipated profit
52.2	performance to continue during arbitration	additional costs occasioned by any suspension ordered during arbitration proceedings

Clause 5.7 – unexpected site conditions

In the case of work underground or involving excavation where the actual conditions of the ground are not stated in the Contract or if rock, rocky soil, solid chalk, water, running sand, slag, pipes, concrete or other obstructions are found, and such conditions or obstructions could not reasonably have been ascertained from an inspection of the Site by the Contractor before he prepared the Tender or from information made available to the Contractor for the purposes of the Tender, or if it should be necessary to leave in timber or provide support for existing work (such necessity not having been indicated in the Contract), the Contractor shall forthwith inform the Engineer of the obstructions or hazards encountered and obtain the approval of the Engineer to the steps he proposes to take to deal with the same. If the Contractor in taking such steps incurs extra Costs, such Costs shall be added to the Contract Price.

The test for entitlement under this clause is not as stringent as in many other contracts with a similar provision. This is an 'unexpected' conditions clause not an 'unforeseen' conditions clause. So providing the contractor can demonstrate that the conditions encountered were neither reasonably ascertainable from site inspection nor indicated in the contract or tender information the entitlement to payment of extra costs arises.

On strict interpretation of the wording of the clause the entitlement is not retrospective and only applies to extra costs incurred subsequent to the contractor obtaining the approval of the engineer to the proposed steps. This may seem a little harsh and it raises the question of what happens if the engineer delays in giving approval or refuses or otherwise avoids giving approval.

The answer is probably that the contractor has no entitlement for extra costs incurred prior to giving notice of encountering the unexpected conditions but the extra costs after such notice could be the subject of a dispute in arbitration. In some circumstances it could be appropriate for the contractor to raise the matter first under clause 2.6.

Note that there is no provision under clause 5.7 for the contract price to be reduced if the contractor encounters better conditions than expected. However, in appropriate circumstances the engineer could possibly deal with this by ordering a variation which could be valued under clause 27.3.

Clause 6.1 – statutory and other regulations

If the Cost to the Contractor of performing his obligations under the Contract shall be increased or reduced by reason of the making after the date of the Tender of any law or of any order, regulation or by-law having the force of law that shall affect the Contractor in the performance of his obligations, the amount of such increase or reduction shall be added to or deducted from the Contract Price as the case may be.

This clause places all the risk of cost increases or decreases resulting from changes in legislation, etc on the purchaser. In some other forms of contract only labour tax matters are at the purchaser's risk and the contractor takes the risk of costs otherwise arising from statutory changes.

Note that it is the date of the tender which is the relevant date for the application of clause 6.1. Thus although pending changes may be known about before the date of the tender such changes nevertheless still come within the scope of the clause.

Clause 6.2 – labour, materials and transport

> If, by reason of any rise or fall in the Cost of labour or in the Cost of material or transport above or below such Costs ruling at the date of the Tender, the Cost to the Contractor of performing his obligations under the Contract shall be increased or reduced, the amount of such increase or reduction shall be added to or deducted from the Contract Price as the case may be. No account shall be taken of any amount by which any Cost incurred by the Contractor has been increased by the default or negligence of the Contractor.
>
> This Sub-Clause shall apply unless specifically excluded by the Special Conditions. Where this Sub-Clause does apply then unless otherwise agreed increases or decreases shall be calculated and determined by reference to a formula to be specified in the Special Conditions.

Few contracts other than MF/1 and MF/2 retain a price fluctuations clause which applies unless specifically excluded. Normally such clauses apply only when specifically included – as with G/90.

The provision that no account is to be taken of cost increases attributable to the default or negligence of the contractor is presumably intended to ensure that the purchaser does not become liable for extra costs as a result of the contractor's culpable delay. It is possible to read the provision as meaning the opposite.

Clause 11.8 – breach of the purchaser's general obligations

> In the event that the Purchaser shall be in breach of any of his general obligations imposed by this Clause then the additional Cost reasonably incurred by the Contractor in consequence thereof shall be added to the Contract Price.

The general obligations imposed by clause 11 are:

- 11.1   access to site
- 11.2   wayleaves and consents
- 11.3   import permits, licences and duties
- 11.4   foundations and buildings
- 11.5   purchaser's lifting equipment
- 11.6   utilities and power
- 11.7   power for tests on site.

Because of the exclusion of the contractor's common law rights of claim, clause 11.8 is essential in providing the contractor with a remedy for breach of obligation by the purchaser. The application of the clause, however, is restricted to the events specified in clause 11 and it does nothing to protect the contractor in respect of other breaches where the contract provides no remedy. If anything the clause acts against the contractor in such situations by expressly indicating the limitation of the remedy for purchaser's breach.

The reference to cost 'reasonably incurred' may be taken as a requirement on the contractor to mitigate. A distinction can be seen here between the entitlement of the contractor to recover 'extra costs' when these arise from instructions and the like and 'reasonable' extra costs when these arise from the purchaser's breach.

Clause 14.5 – revision of programme

> If the Engineer decides that progress does not match the Programme, he may order the Contractor to revise the Programme. The Contractor shall thereafter revise the Programme to show the modifications necessary to ensure completion of the Works within the Time for Completion.
>
> If modifications are required for reasons for which the Contractor is not responsible, the Cost of producing the revised Programme shall be added to the Contract Price.

The contractor's entitlement under clause 14.5 to additional payment can clearly arise from delay or disruption for which the purchaser is responsible or from an order to mitigate the consequences of delay under clause 33.3. It is less clear whether the entitlement arises in respect of disturbance to the programme by matters beyond the reasonable control of the contractor. MF/1 makes the contractor responsible for such matters on costs but not on time. The contractor certainly has an argument that if he gets an extension of time for a matter beyond his reasonable control then he should be paid the costs of producing any revised programme required by the engineer.

The amount of cost recoverable under clause 14.5 is only that for producing the programme itself and the actual costs of delay, disruption or mitigation must be recovered elsewhere.

Clause 16.2 – errors in drawings supplied by the purchaser or the engineer

> The Purchaser shall be responsible for errors, omissions or discrepancies in drawings and written information supplied by him or by the Engineer. The Purchaser shall at his own expense carry out any alterations or remedial work necessitated by such errors, omissions or discrepancies for which he is responsible or pay the Contractor the Cost incurred by the Contractor in carrying out in accordance with the Engineer's instructions any such alterations or remedial work so necessitated.

The apparent contradiction between the purchaser's responsibility for information under this clause and the transferred responsibility under clause 13.3 is discussed elsewhere in this book; as also is the limited scope for the purchaser to carry out himself any alterations or remedial work.

The costs recoverable by the contractor in complying with the engineer's instructions can, it is suggested, include for any delay and disruption costs arising from rectification of errors.

Clause 18.4 – opportunities for other contractors

The Contractor shall, in accordance with the Engineer's requirements, afford all reasonable opportunities for carrying out their work to any other contractors employed by the Purchaser and to the workmen of the Purchaser who may be employed in the execution or near the Site of any work not included in the Contract or of any contract which the purchaser may enter into in connection with or ancillary to the Works. If, however, the Contractor shall, on the written request of the Engineer or the Engineer's Representative, make available to any such other contractor or to the Purchaser any Contractor's Equipment or provide any other service of whatsoever nature, the Purchaser shall pay to the Contractor in respect of such use or service such sum or sums as shall, in the opinion of the Engineer, be reasonable.

This clause anticipates that the contractor will not have sole possession of the site and that some measure of co-operation is required between the various contractors on the site. Similar provisions in other standard forms refer to the contractor affording 'reasonable facilities'.

The contractor's entitlement to additional payment is expressed only in terms of equipment and services and does not extend to delay and disruption costs caused by the presence of other contractors. The only obvious financial remedy for the contractor in MF/1 for such delay and disruption is through clause 25 (suspension of work). That clause does expressly include for prevention by other contractors.

Clause 18.4, it should be noted, is not written as a 'cost' claim. The contractor's entitlement is expressed in terms of 'such sums as may be reasonable'. Consequently, although clause 18.4 claims are not listed under clause 41.2 as claims which attract profit that is not to say that profit cannot be included in the 'reasonable' sums.

Clause 21.2 – special loads

Should the Contractor consider that the moving of one or more loads of Plant or Contractor's Equipment is likely to damage any highway or bridge unless special protection or strengthening is carried out, then the Contractor shall before moving the load notify the Engineer. The Contractor shall in the notice state the weight and other particulars of the load to be moved and his proposal for protecting or strengthening the highway or bridge.

Unless within fourteen days of receipt of such notice the Engineer by notice directs that such protection or strengthening is unnecessary, then the Contractor shall carry out his proposals with any modification thereof that the Engineer may require.

If there are items in the rates specified in the schedule of prices applicable to such work the Contractor shall be paid for such work at the rates specified. If there are not such rates, the Cost of the work shall be certified by the Engineer and added to the Contract Price.

The practical application of this clause is dealt with elsewhere in this book.

The point to make here is that as it stands the clause places the financial risks relating to roads and bridges of getting special loads, whether plant or equipment, to the site on the purchaser.

Clause 22.1 – setting out

The Contractor shall accurately set out the Works in relation to original points, lines and levels of reference given by the Engineer in writing and provide all necessary instruments, appliances and labour therefor.

If, at any time during the execution of the Works, any error appears in the positions, levels, dimensions or alignment of the Works, the Contractor shall rectify the error.

The Contractor shall bear the Cost of rectifying the error, unless the error results from incorrect information supplied in writing by the Purchaser, the Engineer or the Engineer's Representative, or from default by another contractor, in which case the Cost incurred by the Contractor shall be added to the Contract Price.

The Contractor shall indemnify, protect and preserve bench marks, sight rails, pegs and other things used in setting out the Works.

The effect of this clause is that the contractor is only responsible for his own setting out work but it is also clear that the contractor cannot rely on information given orally by the engineer or his representative.

The reference to default by another contractor supplements the provisions of clause 11.4 relating to foundations so that the contractor is not liable for work built out of position by others.

The amount recoverable by the contractor can include for delay and disruption costs as well as the direct costs of rectification.

Clause 25.2 – additional cost caused by suspension

Any additional Cost incurred by the Contractor in complying with the provisions of and/or the Engineer's instructions under Sub-Clause 25.1 (Instructions to Suspend) shall be added to the Contract Price.

Clause 25.1 deals with prevention of progress in the delivery or erection of plant by the engineer, purchaser or some other contractor employed by the purchaser. The clause operates on both actual instructions and deemed instructions and this follows through into clause 25.2. Thus the contractor does not need a written instruction from the engineer to found a claim under clause 25.2.

Because of the limited provisions elsewhere in MF/1 for delay and disruption claims and the exclusion of common law breach claims, the potentially wide scope of clause 25.2 is most important to contractors who suffer interference in their progress.

Other claim clauses

The remaining claim clauses of MF/1 are of a detailed nature and are covered in commentary in the following chapters:

Clause 5.4	(provisional sums)	– Chapter 18
Clause 5.5	(prime cost items)	– Chapter 18
Clause 19.1	(hours of work)	– Chapter 6
Clause 25.5	(resumption of work after suspension)	– Chapter 10
Clause 25.6	(effects of suspension on defects liability)	– Chapter 10
Clause 27.3	(valuation of variations)	– Chapter 16
Clause 27.5	(alterations due to variations)	– Chapter 16
Clause 31.2	(tests during defects liability period)	– Chapter 9
Clause 33.3	(mitigation of the consequences of delay)	– Chapter 11
Clause 36.8	(searches for defects)	– Chapter 14
Clause 40.2	(delayed payment)	– Chapter 18
Clause 40.3	(failure to certify or to pay)	– Chapter 18
Clause 43.3	(damage caused by purchaser's risks)	– Chapter 15
Clause 46.4	(termination for *force majeure*)	– Chapter 19
Clause 51.3	(limitation due to purchaser's default)	– Chapter 19
Clause 52.2	(performance to continue during arbitration)	– Chapter 20

Allowance for profit on claims

This list of clauses under clause 41.2 where the contractor is entitled to be paid an allowance for profit on claims is given in section 17.3 above.

The percentage to be added as profit is not left to the contractor to argue but is fixed as the percentage stated in the appendix. This percentage, which is usually in the region of $2\frac{1}{2}\%$ to 5% is inserted by the purchaser.

17.5 *Claims procedure under MF/1*

One of the major changes from the old MF 'A' to MF/1 was the inclusion in MF/1 of a detailed claims procedure.

Clause 41.1 – notification of claims

> In every case where by virtue of these Conditions circumstances arise which the Contractor considers entitle him to claim additional payment the following provisions shall take effect:
>
> (a) within 30 days of the said circumstances arising the Contractor shall, if he intends to make any claim for additional payment, give to the Engineer notice of his intention to make a claim and shall state the reasons by virtue of which he considers that he is entitled thereto;
> (b) as soon as reasonably practicable after the date of the notice given by the Contractor of his intention to make a claim for additional payment, and not later than the expiry of the last Defects Liability Period, the Contractor shall submit to the Engineer (with copies for transmission to the Purchaser) full particulars of and the actual amount of his claim. The Contractor shall thereafter promptly submit such further particulars as the Engineer may reasonably require to assess the value of the claim.

This clause provides a strict regime which must be followed if the contractor is to secure his entitlement to additional payment. The purpose is to ensure that the purchaser is kept informed of his potential financial liabilities and is not surprised by a late claim.

The phrase 'by virtue of these Conditions' in the first sentence of the clause is inappropriately placed and it is understood that it is likely to be repositioned in the 1995 Revision 3 to MF/1 so as to apply to the entitlement to claim and not to the cause of the claim.

Clause 41.1.(a) has two distinct requirements – notice of intention to claim and reasons for claiming.

Notice

On notice, the contractor has only 30 days (and these it should be noted from the definition in clause 1.1.bb are calendar days not working days) of the claimable event arising in which to inform the engineer of his intention to claim. However, claimable circumstances sometimes develop and extend over a period of time rather than start and finish at a single identifiable event. The claimable circumstances are ongoing and it is probable that the 30 day notice period does no more than fix the date from which recovery of extra costs can be claimed. That is to say 30 days retrospective costs from the date of the notice are the maximum claimable.

Reasons

On reasons, the obligation on the contractor is to identify as best he can the provisions in the contract which entitle him to claim. There is nothing to prevent the contractor giving alternative reasons if the circumstances are appropriate and in the event of uncertainty it is best to do so.

The contractor's claim notice will not be invalidated if he fails to quote the correct reasons for his claim in the first instance – and there appears to be nothing to stop reconsideration of the reasons. However, the claim notice may be invalidated if the contractor fails to state any reasons.

In the case of *Humber Oils Terminal* v. *Hersent Offshore* (1981) in a claim under the ICE Fourth edition Conditions of Contract it was held that a notice of claim was invalid by failing to meet all three specified notice requirements in giving:

- the conditions encountered
- the contractor's proposals
- the anticipated delay.

Particulars

Clause 41.1.(b) allows the contractor some flexibility in the submission of particulars and the amounts of claim. The requirement is expressed 'as soon as reasonably practicable' subject to cut-off at the expiry of the last defects liability period. There is no requirement for interim statements of claim and the words 'full particulars of and the actual amount of' suggest that the contractor can defer the submission of particulars and amounts until he is satisfied that all the costs involved have been recognised and accounted for.

Unlike some standard forms MF/1 does not expressly require the contractor to keep contemporaneous records of claims and to make them available for inspection by the engineer. Clause 41.1.(b) simply requires the contractor to provide such further particulars as the engineer may reasonably require.

Clause 41.3 – purchaser's liability to pay claims

Notwithstanding anything contained in these Conditions the Purchaser shall not be liable to make payment in respect of any claim for an additional payment unless the Contractor has complied with the requirements of this Clause.

The effect of this clause, coupled with clause 44.4 (exclusive remedies) is that the contractor has no rights of claim except where proper notice has been given.

17.6 Claims under MF/2 and MF/3

MF/2

The claims procedure in MF/2 as set out in clauses 29.1 and 29.3 is identical to that in clauses 41.1 and 41.3 of MF/1.

The list of clauses in clause 29.2 on which profit is allowable is smaller than the list in clause 41.2 of MF/1 because MF/2 does not contain provisions relating to the following:

- unexpected site conditions
- special loads
- setting out
- tests during defects liability period
- contractor to search.

Other differences are that the breach of the purchaser's general obligations relates only to permits, licences and duties; and profit is allowable on claims for the effect of suspension on defects liability (but see the note earlier under section 17.3).

In clause 32.4, MF/2 has an identical exclusive remedies clause to that in clause 44.4 of MF/1.

MF/3

MF/3 has no specific claims procedure and more importantly, perhaps, it does not have an exclusive remedies clause. Common law claims are therefore permitted in addition to, or as an alternative to, contractual claims.

However, as would be expected in a supply of goods contract, the vendor does not have a wide range of circumstances giving rise to rights to additional payment.

In MF/3, the express entitlements to additional payment are:

- clause 2.1 (information)   – the purchaser to pay the reasonable extra costs caused by delay in providing information or by the supply of inaccurate information
- clause 8.1 (storage)   – the purchaser to pay the reasonable cost of storing, protecting, preserving and insuring goods when delivery is delayed on the purchaser's instructions
- clause 13.2 (valuation of variations)   – an amount to be added or deducted from the contract price

- clause 16.1 (statutory   – the amount of any increase or decrease in
  and other regulations)   cost caused by changes in legislation, etc
                                to be added or deducted from the contract
                                price.

MF/3 does not define 'cost' nor does it make any reference to allowance
for profit. Consequently, unlike MF/1 and MF/2, the question of whether
profit (or loss of profit) can be included in a claim is best determined from
common law principles.

Note that MF/3 does not contain a price fluctuation clause.

17.7 Claims under G/90

G/90 does not have an exclusive remedies clause and claims for breach
can apply as well as contractual remedies. Consequently whenever in
G/90 the purchaser has an obligation and there is breach of that obligation the contractor, if he suffers damage as a result of the breach, will be
able to look to either a contractual remedy or a common law remedy. And,
for common law claims, the contractor is not limited to breaches of the
express terms of the contractor – claims may also be founded on implied
terms.

Express provisions for additional payment

The following clauses of G/90 deal with adjustments to the contract price
or otherwise express an entitlement to additional payment (or deduction):

Table 17.2 Express provisions for additional payment

Clause	Description	Entitlement
3(i)	security for performance	the expenses of the bond to be paid by the purchaser
3(iii)	contract agreement	the expenses of the contract agreement to be paid by the purchaser
5(ii)	mistakes in information	the purchaser shall pay the extra cost occasioned by errors, omissions, etc in drawings supplied in writing by the purchaser or the engineer
10(i)	variations and omissions	an amount to be added or deducted from the contract price using rates if applicable but otherwise such sum as is reasonable

Table 17.2 Contd

Clause	Description	Entitlement
11	underground works	extra cost determined in like manner to the valuation of variations
12	contractor's default	the contractor to PAY the excess between the cost of completing the works and any amounts due to the contractor
15(i)	delivery of plant	a reasonable sum determined in accordance with clause 10 (variations) to be added to the contract price for additional movements or storage of plant required by the engineer
15(iv)	delayed plant	a sum determined in like manner to the valuation of variations to be added to the contract price for storing, protecting, preserving and insuring delayed plant
15(vii)	examination and making good of delayed plant	a reasonable sum to be added to the contract price for examining and making good, or delivering and erecting delayed plant
16(v)	work outside normal working hours	a sum to be determined in like manner to the valuation of variations to be added to the contract price for work done outside normal working hours on the instruction of the engineer
21(ii)	loss or damage to the works due to excepted risks	a sum determined in like manner to the valuation of variations to be added to the contract price for work required by the purchaser in making good loss or damage caused by the excepted risks
27(v)	repetition of tests before completion	reasonable expenses of the purchaser to be DEDUCTED from the contract price
27(vii)	performance tests	the contractor to PAY liquidated damages for the failure of performance tests
28(iv)	interference with tests before completion	the purchaser to pay a sum determined in like manner to the valuation of variations for prevention of tests by reason of an act or omission of the purchaser, the engineer or a contractor employed by the purchaser

Table 17.2 Contd

Clause	Description	Entitlement
28(v)	protection of the works after deemed taking-over	a sum determined in like manner to the valuation of variations to be added to the contract price for measures taken by the contractor to protect the works after deemed taking-over
28(vi)	delayed tests before completion	sums to be determined in like manner to the value of variations for: 1. examining and making good defects 2. complying with obligations on defects later than stated due to delayed tests on completion to be added to the contract price
29(i)	suspension of works or prevention or delay to progress	a sum to be determined in like manner to the valuation of variations to be added to the contract price for suspension; or for prevention or delay by the purchaser, the engineer or some other contractor employed by the purchaser
32(i)	provisional sums	not stated
32(ii)	prime cost items	the net amount paid by the contractor plus the percentage stated in the appendix
34(iii)	interest on delayed payment	on payments improperly delayed by the purchaser or the engineer interest is payable at 2% over the base rate of the bank named in the appendix
35(i)	statutory and other regulations	the amount of any increase or decrease due to changes in legislation, etc to be added or deducted from the contract price
35(ii)	value added tax	the purchaser to pay value added tax as an addition to other payments due under the contract
39(i)	variation in costs	applies only when the contract is let on a fluctuating price basis.

Clause 5(ii) – mistakes in information

> The Purchaser shall be responsible for, and shall pay the extra cost, if any, occasioned by any discrepancies, errors, or omissions in the drawings and information supplied in writing by the Purchaser or the Engineer.

This is one of the few express entitlements to additional payment in G/90 which does not require the amount due to be calculated in like manner to the valuation of variations.

Cost is not a defined term in G/90 and it is arguable that the phrase 'extra cost' as used in clause 5(ii) should include not only for direct costs and overheads but also for financing charges and profit. Delay and disruption costs should be allowable.

The obvious application of clause 5(ii) is to mistakes in information discovered prior to completion which involve the contractor in extra cost. But the clause may also be seen as providing an indemnity to the contractor in respect of mistakes in information which come to light after completion. There is however some potential conflict in this with the wording of clause 30(i) – (defects after taking over) – in that this latter clause makes the contractor responsible for 'design' furnished or specified by the purchaser unless the contractor disclaims responsibility in writing within a reasonable time of its receipt. To resolve the conflict it seems to be necessary to distinguish between 'drawings and information' and 'design'.

Clause 10(i) – valuation of variations

> The amount to be added to or deducted from the Contract Price shall be ascertained and determined in accordance with the rates specified in the schedules of prices, so far as the same may be applicable, and where rates are not contained in the said schedules, or are not applicable, such amount shall be such sum as is reasonable in the circumstances. Due account shall be taken of any partial execution of the Works which is rendered useless by any such valuation.

The general policy of G/90 is to relate the contractor's entitlement to additional payments to the valuation of variations under the rules of clause 10(i).

The question of whether the valuation is then made on rates which are 'applicable' or is determined as 'such sum as is reasonable' depends to a great extent on how the schedule of prices is formulated. If the schedule is composed of work items (an activity schedule) it will rarely be 'applicable'. But if, as is common, the schedule is of unit labour and plant rates then it will be widely 'applicable'. In this latter situation the contractor is effectively able to price his claims on a daywork basis.

Where 'such sum as is reasonable' is the basis of payment, contractors usually argue that cost-plus applies – and are usually successful in doing so. However, following the case of *Laserbore Ltd* v. *Morrison Biggs Wall Ltd* (1993) discussed in Chapter 16 some contractors may seek to argue that schedules of dayworks should be used.

In the Laserbore case it was held that a 'fair and reasonable payment' was better achieved by using such schedules than by cost-plus.

Clause 11 – underground works

In the case of work underground or involving excavation where the actual conditions of the ground are not stated in the Contract and could not reasonably have been inferred from an inspection of the Site by the Contractor before he prepared his tender, if rock, rocky soil, solid chalk, water, running sand, slag, pipes, concrete, or other obstructions are found, or if it should be necessary to leave in timber or provide support for existing works (such necessity not having been indicated in the Contract), the Contractor shall inform the Engineer as soon as reasonably practicable of the steps he proposes to take to deal with the hazard. If, as a consequence thereof, extra cost is incurred by the Contractor, a sum ascertained and determined in like manner to the valuation of variations under Clause 10 (Variations and Omissions) shall be added to the Contract Price.

This clause is very similar to clause 5.7 of MF/1 – thus many of the comments on that clause in Section 17.4 above apply.

However there are two notable points of difference:

- Clause 11 of G/90 simply requires the contractor to inform the engineer of the steps he proposes to take – the contractor is not required to obtain the approval of the engineer as in MF/1.
- The basis of payment is linked to the valuation of variations rather than being expressed as simply extra cost.

Clause 29(1) – suspension of works

If by reason of the suspension of the Works by the Purchaser or the Engineer (otherwise than in consequence of some default on the part of the Contractor) or by reason of the Contractor being prevented from or delayed in proceeding with the Works by the Purchaser, the Engineer, or some other contractor employed by the Purchaser, the Contractor shall incur additional expense, there shall be added to the Contract Price a sum in respect thereof, such sum to be ascertained and determined in like manner to the valuation of variations under Clause 10 (Variations and Omissions). Provided that no claim shall be made under this clause unless the Contractor has, within a reasonable time after the event giving rise to the claim given notice in writing to the Engineer of his intention to make such claim.

This clause is potentially the most important claims provision in G/90.

The marginal note 'suspension of works' suggests that the clause is of limited effect but the wording of the clause clearly deals with prevention and delay generally. And since it is stated in clause 38 that marginal notes shall not affect the construction of the contract it is the wider application of the clause which applies.

The clause can be seen therefore as giving the contractor entitlement to claim for any prevention or delay which is caused by either:

- suspension of the works
- the purchaser or the engineer
- some other contractor employed by the purchaser.

The effect is that under G/90 the contractor has contractual rights of claim for prevention and delay as well as his common law rights.

The remedy under clause 29(1) which refers to the valuation of variations might produce a different financial result than a common law damages claim. But whether or not there are financial advantages to be gained one way or the other the contractor undoubtedly benefits from having the contractual entitlement.

Firstly, there is no need in a claim under clause 29(1) for the contractor to argue – as would be necessary with a common law claim – that the contract has an implied term that the purchaser accepts responsibility for prevention or delay caused by his engineer or by his other contractors. The clause covers the point.

Secondly, if a suspension is ordered to cover outside interference for which the purchaser would not be legally liable, the contractor can still make a claim under clause 29(1). The clause excludes only default on the part of the contractor from its payment provisions.

Thirdly, the generality of clause 29(1) covers gaps elsewhere in G/90 on the contractor's rights of claim. For example, nothing is said in clause 16 (access to and possession of the site) relating to failure to give access or possession. Without the provisions of clause 29(1) this would be a serious omission.

There is in clause 29(1) a requirement to give written notice to the engineer within a reasonable time after the event giving rise to the claim – a requirement not used consistently throughout G/90. But the requirement is not onerous and is satisfied simply by giving notice of intention to claim.

Clause 35(1) – statutory and other regulations

If the cost to the Contractor of performing his obligations under the Contract shall be increased or reduced by reason of the making or amendment after the date of tender of any law or of any order, regulation, or by-law having the force of law in the United Kingdom that shall affect the Contractor in the performance of his obligations under the Contract, the amount of such increase or reduction (to the extent that it arises directly in respect of the Works) shall be added to or deducted from the Contract Price as the case may be.

This clause is much the same as clause 6.1 of MF/1 but in G/90 it is specifically changed to laws having force in the United Kingdom which are covered.

The phrase in brackets 'to the extent that it arises directly in respect of the Works' is not found in MF/1 and neither its meaning nor its purpose is exactly clear. The phrase seems to be aimed at excluding any costs to the contractor of performing his obligations under the contract arising from legislation changes when such costs are either indirect or consequential. It could perhaps be applied to certain types of overhead costs but it is doubtful if it is intended to exclude claims being passed on by sub-contractors.

Clause 39(1) – variation in costs

If the Contract is let on a fluctuating price basis and if the cost to the Contractor of performing his obligations under the Contract shall be increased or reduced by reason of any rise or fall in the rates of wages payable to labour or in the cost of material or transport above or below such rates and costs ruling at the date of tender, the amount of such increase or reduction shall be added to or deducted from the Contract Price as the case may be, provided that no account shall be taken of any amount by which any cost incurred by the Contractor has been increased by the default or negligence of the Contractor.

The operation of this clause is clearly provisional on a statement elsewhere in the contract that the contract is let on a fluctuating basis. This is the reverse of the position in MF/1 where the corresponding clause 6.2 has to be specifically excluded.

Other claims clauses

These are dealt with in detail in the following chapters:

clause 15(i)	delivery of plant	– Chapter 10
clause 15(iv)	delayed plant	– Chapter 10
clause 15(viii)	delayed plant	– Chapter 10
clause 16(v)	work outside normal hours	– Chapter 6
clause 21 (ii)	loss or damage due to excepted risks	– Chapter 15
clause 28(iv)	interference with tests	– Chapter 9
clause 28(v)	protection of the works after taking over	– Chapter 12
clause 28(vi)	delayed tests	– Chapter 12

clause 32(i)	provisional sums	– Chapter 18
clause 32(ii)	prime cost items	– Chapter 18
clause 34(iii)	interest on delayed payment	– Chapter 18
clause 35(ii)	value added tax	– Chapter 18

17.8 Common law claims

Common law claims fall into various categories:

- claims for breach of an express or an implied term of the contract
- claims for misrepresentation
- claims for *quantum meruit*
- claims for negligence (tort).

Claims for breach

To establish a successful claim for breach of contract it is necessary to prove that damage (in a financial sense) has been suffered as a result of the breach.

The fundamentals of such a claim are briefly:

- breach of contract must be proved
- damage must be proved
- the damage must be shown to flow from the breach (causation)
- the damage must not be too remote.

Measure of damages

The principles applied by the courts in measuring damages date back to the case of *Robinson* v. *Harman* (1848) where it was stated:

'The rule of common law is that where a party sustains a loss by reason of a breach of contract, he is, so far as money can do it, to be placed in the same situation, with respect to damages, as if the contract had been performed.'

But this rule is subject to the rules on remoteness of damage.

Remoteness of damage

The law does not allow a claimant to succeed in every case where damage follows breach but draws a practical line by excluding that which is too remote.

The guiding principles of remoteness applying to cases of breach of contract derive from the judgment of Baron Alderson in the very old case of *Hadley* v. *Baxendale* (1854). The rule in *Hadley* v. *Baxendale* is taken as having two branches and is commonly expressed as:

'Such losses as may fairly and reasonably be considered as either arising:
    (1st rule) "naturally", i.e. according to the usual course of things, or
    (2nd rule) "such as may reasonably be supposed to be in the contemplation of both parties at the time they made the contract, as the probable result of breach of it".'

The test for remoteness laid down by Baron Alderson was reformulated in the classic judgment of Lord Justice Asquith in the case of *Victoria Laundry (Windsor) Ltd* v. *Newman Industries Ltd* (1949) in terms of foreseeability.

However, in *Czarnikow* v. *Koufos* (1969) known as *The Heron II* the House of Lords moved away from the foreseeability test to one of assumed common knowledge. The effect of this on the law and a summary of the law as it now stands was admirably expressed by the Court of Appeal of New Zealand in *Bevan Investments* v. *Blackhall and Struthers* (1978) as follows:

'(1) The aggrieved party is only entitled to recover such part of the loss actually resulting as may fairly and reasonably be considered as arising naturally, that is according to the usual course of things, from the breach of the contract.
(2) The question is to be judged as at the time of the contract.
(3) In order to make the contract-breaker liable it is not necessary that he should actually have asked himself what loss was liable to result from a breach of the kind which subsequently occurred. It suffices that if he had considered the question he would as a reasonable man have concluded that the loss in question was "liable to result".
(4) The words "liable to result" should be read in the sense conveyed by the expressions "a serious possibility" and "a real danger".'

The rules of remoteness it should perhaps be emphasized apply to the type of loss and not to the amount of loss.

Mitigation of loss

It is frequently said that a claimant has a duty to mitigate his loss. This is true only to the extent that the claimant seeks to recover his loss as damages. It does not follow that an injured party in a breach of contract should have his conduct determined by the breach.

The following extracts from legal judgments explain this. Viscount Haldane in *British Westinghouse Electric and Manufacturing Co Ltd* v. *Underground Electric Railways of London Ltd* (1912) said:

> 'A plaintiff is under no duty to mitigate his loss, despite the habitual use by the lawyers of the phrase "duty to mitigate". He is completely free to act as he judges to be in his best interest. On the other hand, a defendant is not liable for all loss suffered by the plaintiff in consequence of his so acting. A defendant is only liable for such part of the plaintiff's loss as is properly to be regarded as caused by the defendant's breach of duty'.

Sir John Donaldson, Master of the Rolls, in *The Solholt* (1983) said:

> 'The fundamental basis is thus compensation for pecuniary loss naturally flowing from the breach; but this first principle is qualified by a second, which imposes on a plaintiff the duty of taking all reasonable steps to mitigate the loss consequent on the breach, and debars him from claiming any part of the damage which is due to his neglect to take such steps.'

Causation and global claims

Strict application of the rules of causation and the burden of proving loss create serious difficulties in claims made under plant, process and construction contracts. Frequently the contractor finds it difficult to identify how much of his losses are his own responsibility and how much should be charged to the purchaser; and that is before he starts the task of trying to allocate his costs to individual claimable events.

Claiming on a global basis obviously alleviates these difficulties; but the question is how acceptable is it for the contractor to roll-up his costs and put them forward as the composite amount for a bundle of claims?

The cases of *J. Crosby & Sons Ltd* v. *Portland Urban District Council* (1967) and *London Borough of Merton* v. *Stanley Hugh Leach* (1985) are often quoted as authority for the practice but they need to be treated with caution. They apply only to quantification and then only when the contractual machinery is no longer effective.

The current legal trend is against global claims and the need for a claimant to link cause and effect and state his case with particularity was confirmed by the Privy Council in the case of *Wharf Properties Ltd* v. *Eric Cumine Associates* (1991) where it was said of the *Crosby* and *Merton* cases:

> 'Those cases establish no more than this, that in cases where the full extent of extra costs incurred through delay depend upon a complex interaction between the consequences of various events, so that it may

be difficult to make an accurate apportionment of the total extra costs, it may be proper for an arbitrator to make individual financial awards in respect of claims which can conveniently be dealt with in isolation and a supplementary award in respect of the financial consequences of the remainder as a composite whole. This has, however, no bearing upon the obligation of a plaintiff to plead his case with such particularity as is sufficient to alert the opposite party to the case which is going to be made against him at the trial.'

Quantum meruit

Claims for *quantum meruit* – meaning 'what it is worth' are sometimes called quasi-contractual claims. They are highly popular with contractors wishing to escape from the rigidity of a lump sum or from contract rates towards payment on a cost-plus basis.

Strictly speaking the phrase '*quantum meruit*' applies to the law of restitution for the value of services rendered where there is no contractual entitlement to payment. But it is also commonly used to describe claims made under a contract for a fair valuation or a reasonable sum. The division between the two is not always clear-cut as this comment by Mr Justice Goff in *British Steel Corporation* v. *Cleveland Bridge & Engineering* (1981) shows:

'a *quantum meruit* claim straddles the boundaries of what we now call contract and restitution; so the mere framing of a claim as a *quantum meruit* claim, or a claim for a reasonable sum, does not assist in classifying the claim as contractual or quasi-contractual.'

In the broad sense, claims for *quantum meruit* can be made where:

- no contract is ever concluded;
- the contract is unenforceable;
- the contract is void for mistake;
- the contract is discharged by frustration;
- the contract is for a lump sum and the purchaser prevents completion;
- no price is fixed in the contract;
- the work undertaken falls outside the scope of the contract;
- the contract provides for payment of a reasonable sum or a fair valuation.

However, the law of restitution is complex and the cases where claims succeed frequently turn on particular facts. Such cases often prove to be of little lasting authority.

Thus the old case of *Bush* v. *Whitehaven Port and Town Trustees* (1888) much quoted by contractors in support of winter working claims was held by the House of Lords in *Davis* v. *Fareham UDC* (1956) not to be authority for any principle of law. Bush, the contractor, was prevented by late possession of the site from commencing work on a pipeline contract until winter when wages were higher and the work was more difficult to construct. The court held that as a summer contract had been contemplated Bush was entitled to recover his extra expenditure as damages or to be paid for all the work on *quantum meruit*.

The proposition that the *Bush* case was said to support was that where the circumstances of a contract are altered the contract price is no longer binding and can be replaced by *quantum meruit*. This has obvious appeal to contractors. But two recent cases show that the courts are not disposed to award *quantum meruit* where there are contractual provisions for compensation.

In *Morrison-Knudsen* v. *British Columbia Hydro and Power Authority* (1985) the contractor could have terminated the contract because of serious breaches by the employer. Instead the contractor elected to continue working and claimed for the value of work on *quantum meruit*. The court held that such an award could not be given as an adequate remedy was available under the contract.

In *McAlpine Humberoak* v. *McDermott International* (1992) the contract for the deck structure of an off-shore drilling rig was let on a lump sum basis. The contractor claimed extra costs arising from variations. The judge at first instance relying, rather surprisingly, on the *Bush* decision, held that the contract had been frustrated by the extent of the variations and awarded the contractor *quantum meruit*. The Court of Appeal overturned this decision and held:

- the variations did not transform the contract or distort its substance or identity;
- the contract provided for variations;
- the contractual machinery for valuing variations had not been displaced;
- an award of *quantum meruit* could not be supported.

Chapter 18

Valuations and payments

18.1 Introduction

This chapter deals with the arrangements in the various model forms for plant contracts for interim and final valuations and payments.

Valuations

Final valuation of the works in plant contracts is usually a straightforward process (even if some of the figures may be in dispute) in that the contract price is normally on a lump sum basis and the contractor is entitled to the contract price subject to additions and deductions for variations and other matters as detailed in the preceding chapter.

Interim valuations present more of a problem because with lump sum contracts the value of the works until they are completed is more a matter of opinion than of fact. The processes which are commonly used in lump sum construction contracts for interim valuations – activity schedules, milestone schemes, and so on – are not necessarily suitable for plant contracts because it is questionable in plant contracts what, if anything, the purchaser has of value until the works are completed.

Consequently, there are no fixed rules for interim valuations constant to all plant contracts. MF/1 and MF/2 contain only generalized fall-back provisions; MF/3 contains nothing; and G/90 contains alternative schemes which require individual finalization for each contract. It is, therefore, essential that every plant contract sets out (usually in the special conditions) its particular scheme for interim valuations.

Payments

In addition to specifying rules on valuations most contracts also specify rules and obligations relating to payments, including:

- application procedures
- certification procedures

- times for payment
- guarantee requirements
- retention provisions
- remedies for late payment
- finality of certificates.

These could, in principle, be standardized across whole ranges of contracts – lump sum, remeasurement, reimbursable – and across all types of
work – plant, process and construction. Perhaps only guarantee
requirements need to be considered on an individual basis. In practice,
however, the various families of contracts each have their own approach
to these matters and even amongst the plant contracts there are significant
differences between the model forms.

In fact with plant contracts, individuality seems to be encouraged
rather than consistency. The Official Commentary on MF/1 says that
clause 40 (payments) is only a fall-back provision to apply in the absence
of other provisions in the special conditions. The effect of this is that
contractors cannot rely on the application of standard terms and they
need to examine each contract in detail to decide whether it provides a fair
basis for doing business.

18.2 *Valuations generally*

The value of the works of a plant contract as viewed by the contractor
from the standpoint of his costs will increase according to the amount of:

- design and drawings undertaken
- receipt of materials
- plant in the course of manufacture
- plant delivered
- work executed on site
- claims for additional payment.

The value of the works as recognized by the purchaser at any particular
stage for the purpose of making payments to the contractor may include
allowances for all or some of the above items or the value may be assessed
wholly on a time-related basis.

The contract does not necessarily set out to bring the values as seen by
the contractor and the purchaser together and one has to look at the
valuation rules to see how closely they coincide. At one extreme, for
example in MF/3, value is not recognized nor payment made until
delivery; but in G/90, by contrast, one of the alternative payment schemes
recognizes value and allows payment, at least in part, for all the listed
items.

For the most part, discrepancies between the parties' views on value have no contractual significance; because providing the valuation rules of the contract are clear they alone should be used to determine value for payment purposes. It may happen, however, that occasionally exceptional circumstances apply – prolonged suspension or termination for *force majeure* – and it is necessary to ascertain a more objective value of the works. This can rightly be called the contract value.

The 'Contract Value'

In MF/1, MF/2 and G/90 a definition is given of the 'Contract Value'. The definition in clause 1.1.i of MF/1 (and MF/2) reads:

> 'Contract Value' means such part of the Contract Price, adjusted to give effect to such additions or deductions as are provided for in the Contract, other than under Sub-Clause 6.2 (Labour, Materials and Transport), as is properly apportionable to the Plant or work in question. In determining Contract Value the state, condition and topographical location of the Plant, the amount of work done and all other relevant circumstances shall be taken into account.

The definition in G/90 is much the same.

None of the definitions state when the 'Contract Value' is to apply – but as a defined term it is only relevant where it is expressly mentioned elsewhere in the contract. As stated above, the use of the defined term is infrequent and related mainly to exceptional circumstances and (in the case of MF/1 and MF/2) to damages for late completion.

Payment schemes

The terminology used to describe the various types of payment schemes is somewhat imprecise but the following descriptions correspond with popular usage:

- Periodic payments. These give the contractor a right to regular payments (usually monthly) with values based on either:
  - the contract price divided equally
  - measurement of work done
  - progress relative to programme based on a pre-determined expenditure plan
  - assessment of work completed against an activity schedule which breaks down the contract price into lesser lump sums.
- Stage payments. These entitle the contractor to payment when he has reached a specified stage in the works – which may be either an event based stage or a time elapsed stage.

● Milestone payments. These combine the characteristics of periodic payments and stage payments. The contractor is entitled to periodic payments but only to the extent that he has completed specified activities.

18.3 Certificates generally

In contracts where there is provision for both interim and final payments there is usually a requirement that the engineer (or other contract administrator) should issue certificates stating the amounts to be paid by the purchaser to the contractor.

The role of the engineer as certifier is discussed in Chapter 8 but, in short, when the contract requires the engineer to exercise his discretion as certifier he is obliged to act fairly between the parties. That is to say when it is left to the engineer to state his opinion of the amount due – either on the basis of the contractor's application or his own valuation – the engineer must form his opinion using his professional judgment and without regard to external pressures.

Certificates in the model forms

MF/1, MF/2 and G/90 all require the engineer to issue interim and final payment certificates (MF/3 does not – there is no engineer; no certification process; and payment merely becomes due on delivery). However MF/1, MF/2 and G/90 are not as clear as some other standard forms in requiring the engineer to certify only the amounts which in his opinion are due.

All three model forms allow the engineer to exercise his discretion in respect of interim certificates. But for final certificates, if the engineer has any discretion, it is left to be implied. This is from a practical, if not from a legal point of view, an odd situation and the reverse would be more understandable. This is because for interim applications (at least under MF/1 and MF/2) the contractor is required to submit an invoice – and adjustment to invoices is not an acceptable business practice whereas for the final application the contractor is required to submit a statement of final account – and this can, of course, be the subject of adjustment.

Finality of certificates

In some contracts, amongst them MF/1 and MF/2, the final payment certificate serves a dual function. It is conclusive evidence not only of the final value of the works but also that the contractor has performed all his obligations under the contract.

In this respect the final payment certificate takes the place of the defects correction certificate, the certificate of making good defects or the acceptance certificate of other contracts. But whether it is really in the interests of either party that a payment certificate should serve this function is open to question.

One of the problems is that once the engineer has issued the final certificate he is, to use a phrase, *functus officio* – having discharged his duty. The only person left with authority to modify the final certificate is an arbitrator appointed under the contract. Usually the time limits on commencing arbitration proceedings after the issue of the final certificate are short – three months in MF/1 and MF/2; 1 month in G/90.

18.4 Payments generally

Interim payments

In the absence of express provisions in a contract for interim payments, a contractor seeking such payment would have to prove an implied term. There is no general legal entitlement to interim payments under statute or at common law.

It is questionable whether the case for an implied term would succeed in plant contracts because of the uncertain value to the purchaser of the plant until it is completed. And it is most unlikely that any implied term would extend to goods and materials off-site.

MF/3 has no provisions for interim payment and – since the model form expressly states (in clause 10.1) that unless otherwise agreed payment becomes due on delivery – the contractor has no right to interim payment. The wording alone is sufficient to exclude the possibility of an implied term.

MF/1, MF/2 and G/90 all envisage interim payments. And all provide fall-back schemes in the event that a contract does not specify its own particular intentions.

Time for payment

It is a source of complaint in much of British industry that payments are not made within specified terms of payment or, where no terms exist, within reasonable times.

Suppliers of goods and services which have been delivered can do little but sue on their unpaid invoices. But contractors with entitlements to interim payments and whose work is not completed are usually in a stronger position – with a variety of practical options, such as withholding delivery in the event of non-payment.

To regulate the position most standard forms of contract stipulate both the times within which the engineer must certify from receipt of the contractor's application and the times within which the employer must pay – either from delivery of the application or the date of the certificate.

In MF/1 and MF/2 the standard time periods for interim payments are:

- certificates within 14 days of application
- payments with 30 days of certification.

But both MF/1 and MF/2 allow the standard times to be varied in the special conditions.

G/90 states 14 days for certification and 14 days for payment – although these times are frequently increased by amendment.

Remedies for late payment

Not all standard forms of contract provide express remedies for failure by the engineer to certify or failure by the purchaser to make payment within the terms of the contract. There are two situations to consider. Should the contractor be entitled to slow down or suspend progress in the event of non-payment? And should the contractor be entitled to claim interest on payments made late?

On the first, in the absence of express provisions, the contractor faces difficult decisions when confronted with delayed payment – either to assume that the breach of contract is sufficient to justify slowing down, suspension or termination or to proceed normally in the hope that payment will be made eventually. Each case will have its particular merits but as a general rule the contractor should not make any assumptions on rights to slow down, suspend or terminate without consulting his lawyers. To take such action even in the response to a breach could well be a breach itself exposing the contractor to liability for damages.

On the matter of interest for late payment there is a legal presumption that in the ordinary course of things loss is not suffered by reason of late payment. This presents a serious obstacle to the recovery of interest in the absence of an express contractual entitlement. In some circumstances it can be overcome by claiming interest as special damages but this again is very much a matter for lawyers.

MF/1, MF/2 and G/90 address both the contractor's remedies for non-payment and his right to interest on late payments. MF/3, however, is silent on both.

Retention monies

In contracts which provide for interim payments before completion and which impose obligations on the contractor to remedy defects occurring within a specified period after completion it is normal to include provisions for money to be retained by the purchaser from the gross amounts due to the contractor.

There has been much debate in the construction industry as to the legal status of monies held as retentions and the extent to which they are held in trust. These are matters of particular interest when an employer (purchaser) becomes insolvent and the contractor tries to lay claim to the retention monies in preference to other creditors. Some building contracts require the employer to hold retention monies in separate accounts on a trust basis and in the case of *Wates Construction (London) Ltd* v. *Franthom Property Ltd* (1991) it was held that even in the absence of an express obligation to place retention money into a separate account the employer was obliged to do so. However it was also held in the case of *MacJordan Construction Ltd* v. *Brookmount Erostin Ltd* (1991) that if the employer's insolvency pre-dates the establishment of a separate retention account then the contractor is in no better position than any ordinary creditor in respect of monies held.

Of the model forms for plant contracts only G/90 expressly provides for retention money but it is usual for special payment provisions applied to MF/1 and MF/2 to specify retention percentages.

The decision in the *MacJordan* case above almost certainly applies to plant contracts but whether the decision in the *Wates* case similarly applies is open to question.

Because there is a greater element of financial risk to the purchaser in plant and process contracts than there is to the employer in construction contracts in respect of interim payments it is common in plant and process contracts, but rare in construction contracts, for such payments to be made conditional upon the contractor supplying bonds and/or guarantees. These are separate from performance bonds as discussed in Chapter 3 and they relate specifically to the repayment of monies advanced by the purchaser prior to completion.

G/90 contains a selection of such bonds and guarantees. There is also provision in G/90 for the release of all retention money on taking-over subject to the contractor supplying a retention repayment bond. Although not expressly provided for in other plant and process model forms, the practice is commonly written into the special payment conditions for such forms.

Provisional sums

Provisional sums are included in contracts to cover various eventualities such as:

- work which is not fully defined
- work which may, or may not, be required
- allowances for dayworks
- general contingencies.

Provisional sums are almost invariably stated to be expendable solely at the discretion of, and on the instructions of, the engineer. Strictly, therefore, it is wrong to describe as provisional sums items for general contingencies or other items which it is never intended that the engineer will instruct on or will never form part of the final contract price.

For the sake of clarity, lump sum contracts should always state that the contract price will be adjusted to take account of the expenditure under provisional sums. Otherwise disputes can develop on whether the stated sums were included in the tender price at the risk of the contractor or the purchaser.

Any provisional sums to be included in tender prices should therefore always be set by the purchaser to avoid the possibility of a tenderer gaining a competitive advantage by deliberately including low sums which will have to be adjusted upwards in the final account.

Not all contracts address the question of how provisional sums are to be valued. Some refer to the rules for the valuation of variations but others, including the model forms for plant contracts, leave the matter open.

As to whether it is a breach of contract for the purchaser to omit the work covered by a provisional sum and give it to another contractor much depends on the wording of the particular contract. But in principle the contractor should be entitled to claim the right to undertake all the work which he is obliged to carry out, if and when instructed.

Prime cost items

Prime cost items differ from provisional sums in vital respects:

- generally the definition of prime cost fixes the method of valuation – namely, cost-plus;
- in some contracts prime cost items, and only prime cost items, cover nominated sub-contractors and nominated suppliers.

In the event of the purchaser omitting a prime cost sum to give the work to another contractor, that would, it is suggested, be a breach entitling the contractor to claim as damages the percentage mark up he would otherwise have collected.

Fluctuation in price clauses

Contractual provisions which permit a lump sum price or tendered rates to fluctuate according to general rises and falls in commodity and labour costs have various names:

- fluctuation clauses
- variation in price clauses
- contract price adjustments
- inflation provisions.

Without such provisions the contractor takes the financial risks of price movements and is deemed to have allowed for them in his tender. Where the provisions are included in the contract the purchaser takes the risk and some of the price certainty of lump sums and tender rates is inevitably lost.

In times of low inflation there appears to be little need for such clauses but they remain in MF/1 and MF/2 unless specifically excluded. In G/90 the fluctuations clause applies only when specifically included. Surprisingly perhaps, in the light of this, it is G/90 which includes formulae for calculating price changes and not MF/1 (or MF/2) which rely on formulae being stated in the special conditions.

Value added tax

As a general rule prices are deemed to include VAT unless expressly stated otherwise.

Since most contracts require prices to be quoted exclusive of VAT they contain provisions which make it clear that the purchaser is responsible for such VAT as is payable and that accordingly the tendered prices are deemed to exclude VAT. This is the position in MF/3, G/90 and virtually every other standard form of plant, process and construction contract. MF/2 covers the point in the forms of tender and agreement. But surprisingly MF/1 (unlike the old MF 'A') says nothing on VAT. Contractors need to bear this in mind when tendering. However, it is expected that the 1995 Revision 3 to MF/1 will make it clear that the contract price is exclusive of VAT; probably by words to that effect in the forms of tender and agreement.

18.5 *Valuations and payments in MF/1*

Contract price

Clause 1.1.h defines the contract price:

> 'Contract Price' means the sum stated in the Contract as the price payable to the Contractor for the execution of the Works.

This is an unusual definition because the price payable to the contractor for the execution of the works is clearly intended in MF/1 to be the sum stated in the contract adjusted by the various provisions of the contract (see the definition of contract price in MF/2).

As it stands in MF/1, the defined contract price is the tender sum and this is confirmed in the form of agreement which effectively states that the tender sum is 'hereinafter called the Contract Price'. However, the form of agreement does state that the purchaser shall pay the contract price or such other sum as may become payable under the provisions of the contract.

Contract value

Clause 1.1.i defines the contract value:

> 'Contract Value' means such part of the Contract Price, adjusted to give effect to such additions or deductions as are provided for in the Contract, other than under Sub-Clause 6.2 (Labour, Materials and Transport), as is properly apportionable to the Plant or work in question. In determining contract Value the state, condition and topographical location of the Plant, the amount of work done and all other relevant circumstances shall be taken into account.

This is of limited effect in the contract and it applies when there is termination of the contract for prolonged suspension or *force majeure* and in the calculation of liquidated damages for late completion.

Provisional sums

These are defined in clause 5.4:

> Provisional sums included in the Contract Price shall be expended or used as the Engineer may in writing direct and not otherwise. In so far as a provisional sum included in the Contract Price is not expended or used it shall be deducted from the Contract Price.

The definition does not appear to contemplate the possibility that the expenditure under a provisional sum might exceed the amount of the stated sum and that it might be necessary to make additions to the contract price. It could therefore be argued that the amount in a provisional sum is the limit of the contractor's obligation under the provisional sum.

Prime cost items

These are defined in clause 5.5:

All sums included in the Contract Price in respect of prime cost items shall be expended or used as the Engineer may in writing direct and not otherwise. To the net amount paid by the Contractor in respect of each prime cost item there shall be added the percentage thereof stated in the Appendix.

The wording of this clause is again far from precise. It merely implies that the contract price will be adjusted for actual expenditure. And in respect of the percentage additions it has to be assumed that the phrase 'there shall be added' means 'added to the contract price and paid to the contractor'.

Contractor's responsibility for work under provisional sum and prime cost items

Clause 5.6 applies to both clause 5.4 (provisional sums) and clause 5.5 (prime cost items) although it is perhaps intended to apply only to clause 5.5 – see clause 5.6 of MF/2. However, it is understood that the 1995 Revision 3 to MF/1 will bring MF/1 into line with MF/2 by applying clause 5.6 only to clause 5.5.

> The Contractor shall have no responsibility for work done or Plant supplied by any other person in pursuance of directions given by the Engineer under this Clause unless the Contractor shall have approved the person by whom such work is to be done or such Plant is to be supplied and the Plant, if any, to be supplied.

The effects of this provision are exceedingly far reaching. The contractor virtually has an option on whether or not to accept responsibility for work done by nominated sub-contractors and suppliers.

Should such sub-contractors and suppliers prove critical to overall performance then the contractor could be relieved of his own obligations for performance.

Certificate and payment options

Clauses 39 and 40 set out in considerable detail the provisions of MF/1 for applications, certificates and payments. Nevertheless these clauses are intended, at least in part, to operate only as fall-back provisions and it is expected that the special conditions will state in further detail the arrangements for each particular contract. The *aide memoire* for the special conditions printed in MF/1 gives suggested clauses.

The difficulty with this approach is that it can lead to confusion as to which parts of clauses 39 and 40 should be regarded as 'core' provisions –

to be left as standard and not particularised; and as to which parts (if any) of the standard clauses remain in place when additional clauses are stated in the special conditions.

It is suggested that structured analysis of the clauses be made along the following lines:

Table 18.1 Certificate and payment options: structural analysis

Clause number	Description	Core clause to remain unamended in the general conditions	Optional clause to be particularized in the special conditions
39.1	application for payment		✓
39.2	form of application		✓
39.3	issue of payment certificate	✓	time period amendable
39.4	value included in certificates of payment	✓	
39.5	adjustment to certificates	✓	
39.6	corrections to certificates	✓	
39.7	withholding certificates of payment	✓	
39.8	effect of certificates of payment	✓	
39.9	application for final certificate of payment	✓	
39.10	value of final certificate of payment	✓	
39.11	issue of final certificate of payment	✓	time period amendable
39.12	effect of final certificate of payment	✓	
39.13	no effect in the case of gross misconduct	✓	
40.1	terms of payment		✓
40.2	delayed payment	✓	base rates to be stated and type of interest
40.3	remedies on failure to certify or make payments	✓	time periods amendable

Final certificates

By clause 39.9 the contractor 'shall' apply for the final certificate when he has completed his defects liability obligations for the whole of the works and 'may' apply for a final certificate in respect of any section of the works. Any application whether it be for the whole of the works or a section shall be accompanied by a final account stating the full amount of the value of work done and any claims for additional payments.

Since it is not always possible to apportion claims between sections or determine their value until completion of the whole of the works the requirement to include claims in the final account could have the effect of preventing the contractor applying for sections.

The requirement in clause 39.11 for the engineer to issue a final certificate within 30 days of receiving an application is often amended to a longer time period – 60 or 90 days. But, in any event, time only runs from when the contractor has provided all the information reasonably required by the engineer.

The provisions of clause 39.12 that a final certificate is conclusive evidence as to value and performance of the contractor's obligations applies to both sections and the whole of the works. This is a point which needs to be considered carefully by the purchaser (or engineer) in defining sections for practical as well as contractual reasons.

Disputes on final certificates

By clause 39.12 a final certificate become conclusive unless either party commences proceedings either before its issue or within three months thereafter.

Commencing proceedings in this context will normally mean serving notice of arbitration under clause 52.1 but the clause also allows for the commencement of 'other proceedings'. This is presumably intended to cover court action embarked upon as an alternative to arbitration.

Payments

The obligation of the purchaser as stated in clause 40.1 (and usually in any replacement special conditions) is to pay the contractor the amount certified.

Providing therefore that the purchaser does pay the amount certified the contractor would be in difficulty in alleging breach of contract even if the sum certified was patently in error and inadequate. See the case of *Lubenham Fidelities & Investment Co Ltd* v. *South Pembrokeshire District*

Council (1986) where a contractor mistakenly assumed that in this situation he was entitled to terminate the contract.

Delayed payments

Clause 40.2 which gives the contractor entitlement to interest on delayed payments expressly states two important matters:

- interest is payable without formal notice;
- interest is payable without prejudice to other rights and remedies.

Unfortunately it fails to state whether interest is simple or compound. This needs to be clarified in the special conditions. The point on formal notice is that a claim for interest on delayed payment is not caught by the requirements of clause 41 (claims). As to other rights and remedies, the point is that if the contractor takes action under clause 40.3 (remedies on failure to certify or make payment) his entitlement to interest remains intact.

Remedies on failure to certify or make payment

Clause 40.3 applies to two situations:

- when the engineer has failed to issue a certificate, or
- when the purchaser has failed to pay on a certificate.

However the contractor's remedies are not the same for each. Thus the contractor can:

- stop work by giving 14 days notice for either situation, or
- terminate by giving 30 days notice, but only when the engineer fails to issue a certificate.

If the Engineer fails to issue a certificate of payment to which the Contractor is entitled or if the Purchaser fails to make any payment as provided in this Clause the Contractor shall be entitled:

(a) to stop work until the failure be remedied, by giving 14 days notice to the Engineer and the Purchaser in which event the additional Cost to the Contractor occasioned by the stoppage and the subsequent resumption of work shall be added to the Contract Price, and/or
(b) to terminate the Contract by giving 30 days notice to the Engineer and the Purchaser in any case where the Engineer has failed to issue a certificate of

payment, whether or not the Contractor has previously stopped work under paragraph (a) of this Sub-Clause.

The meaning of failing to issue a certificate is, it is suggested, less complex than the meaning of the phrase 'failed to certify' which has troubled the courts in numerous civil engineering cases. Those cases had to consider whether by certifying less than was due to the contractor there had been a failure to certify. In clause 40.3, however, the words are 'fails to issue' which seems reasonably clear.

The remedy of termination stated here applies only when the engineer has failed to issue a certificate. The contractor's remedy of termination for non-payment by the purchaser on a certificate is found in clause 51.1(a). However, it is understood that the 1995 Revision 3 to MF/1 will delete the reference to the engineer failing to issue a certificate from clause 40.3 so the contractor will then be also entitled to terminate for non-payment under clause 40.3.

Note that the contractor has to give notice to both the engineer and the purchaser if he intends to either stop work or terminate. Such notice must, of course, be in writing under the provisions of clause 1.4 (notices and consents).

Retention

The standard provisions of MF/1 say nothing about retention and this should be covered in the special conditions.

18.6 Valuations and payments in MF/2

MF/2, although generally following the text of MF/1, does attend to the various curiosities and omissions in MF/1. Thus:

- Value added tax is mentioned in the form of tender making it clear that the contract price is VAT exclusive.
- The contract price is defined as the sum stated in the contract as adjusted for definitions and variations.
- The contractor can only disclaim responsibility for work done by nominated sub-contractors and suppliers under prime cost items (and not also provisional sums).
- The contractor is entitled to terminate the contract under clause 28.3 for the purchaser's failure to pay as well as for the engineer's failure to issue a certificate.

18.7 Valuations and payments in MF/3

There is no provision in MF/3 for valuation or certification procedures.

Terms of payment

Clause 10.1 states briefly:

> Unless otherwise agreed, payment for the goods shall become due on completion of delivery in accordance with the Contract.

It is therefore for the Vendor to submit his invoice at the appropriate time for the full amount of the goods. There is no provision for retention.

Value added tax

Clause 17.1 applies:

> Unless otherwise stated in the Tender the Contract Price is deemed to exclude Value Added Tax. To the extent that the Tax is properly chargeable on the supply to the Purchaser of any goods or services provided by the Vendor under the Contract, the Purchaser shall pay such Tax as an addition to payments otherwise due to the Vendor under the Contract.

18.8 Valuations and payments in G/90

G/90 contains similar provisions to MF/1 for valuations, certificates and payments but alternative payment schemes are stated within the model form rather than being left for the special conditions.

Contract price

Clause 1 states in its definition of terms:

> The Contract Price shall mean the sum named in the Contract as the Contract Price.

This is effectively the same definition as in MF/1. But the objections of MF/1 do not apply since it is not stated in G/90 that the contract price is the sum to be paid to the contractor for performance of his obligations.

Contract value

This is defined in clause 1 as:

> The 'Contract Value' shall mean that part of the Contract Price, adjusted to give effect to such additions or deductions as are provided for in Clause 10 (Variations and Omissions), which is properly apportionable to the Plant or work in question, having regard to the state, condition and location of the Plant, the amount of work done and all other relevant circumstances but disregarding any changes pursuant to Clause 39 in the cost of executing the Works.

This again serves the same purpose as in MF/1 although its application to liquidated damages for late completion is less certain.

Value added tax

This is covered in clause 35(ii):

> Unless otherwise stated in the tender the Contract Price is deemed to exclude Value Added Tax. To the extent that the Tax is properly chargeable on the supply to the Purchaser of any goods or services provided by the Contractor under the Contract, the Purchaser shall pay such Tax as an addition to payments otherwise due to the Contractor under the Contract.

Provisional sums

Clause 32(i) applies:

> A provisional sum included in the Contract Price shall be expended or used as the Engineer may in writing direct and not otherwise. In so far as a provisional sum is not expended or used it shall be deducted from the Contract Price.

Again this is similar to MF/1.

Prime cost items

Clause 32(ii) applies:

> All sums included in the Contract Price in respect of P.C. (Prime Cost) items shall be expended or used as the Engineer may in writing direct and not otherwise. To the net amount paid by the Contractor in respect of each P.C. item there shall be added the percentage named in the Appendix or the said amount. The sum by which the net amount so paid in respect of any P.C. item plus the

said percentage thereon exceeds or is less than the sum included in the Contract Price in respect of that item shall be added to or deducted from the Contract Price as the case may be.

This definition is an improvement on that in MF/1 in its description of adjustment of the contract price.

Contractor's responsibility

Clause 32(iii) states:

The Contractor shall have no responsibility for work done or Plant supplied by any other person in pursuance of directions given by the Engineer under this clause unless the Contractor shall have approved the person by whom such work is to be done or such Plant is to be supplied and the Plant, if any, to be supplied.

As with MF/1 the words 'under this clause' appear to apply to both provisional sums and to prime cost items. The same objections to this provision as in MF/1 apply.

Alternative payment schemes

Table A of the appendix to G/90 sets out alternative interim payment schemes.

Column 1 of the table is time based and column 2 is event based but the figures are such that on a typical contract the resulting payments are much the same whichever scheme is selected. The time-based scheme of column 1 is intended for use when the number of events makes the use of column 2 impracticable. The purchaser should delete the column which is redundant.

Stage payments

Clause 40 adds detail to interim payment schemes detailed in Table A by setting out provisions in respect of:

- progress relative to programme for time related payments
- completion for event (stage) related payments
- sureties
- vesting of plant
- repayments.

Table A

Clauses 31 and 40: Programme of stage payments for Execution of Contract Works up to Delivery.

At (a) figure between 10 and 15 to be inserted by Purchaser as appropriate to extent of drawings and preliminary work.

At (b) the figure as at (a) less 10% to be inserted by Purchaser.

FIRST Section where sub-divided into two or more Sections as more particularly set out in the Specification.

Column 1*	Column 2*	Column 3**	
Percentage of Delivery Period after which application for payment may be made	Description of Stage reached up to Delivery (more particularly defined in the Specification)	Percentages of Total Price to be paid for the Plant	
		Gross for certification	Net for payment
Approval of drawings to enable civil design to proceed	Approval of drawings to enable civil design to proceed	(a)	(b)
60	Receipt of all major materials, e.g. Castings and Fabrications	40	36 + interim CPA
75	50% of manufacture e.g. Machining of Major Castings	65	$58\frac{1}{2}$ + interim CPA
100	Delivery	100	90 + final CPA on manufacturing phase

By clause 40(i)(a) the contractor's entitlement to stage payments on a time-related basis is qualified if progress is not up to programme and delivery on time is not expected.

By clause 40(i)(b) specified stages (or events) must be completed to the satisfaction of the engineer.

By clause 40(ii) stage payments are subject to the contractor obtaining approved sureties and complying with requirements on the vesting of plant.

By clause 40(iii) if, after receiving stage payments, the contractor fails to deliver on time the purchaser can demand repayment of amounts paid to the contractor.

Interim and final certificates

Clause 31(i) simply states the contractor's rights to apply for payment certificates.

Interim certificates

Clauses 31(ii) to 31(v) deal with interim certificates:

- clause 31(ii) – application for certificates – requires the contractor to substantiate his applications.
- clause 31(iii) – engineer to certify – requires the engineer to certify within 14 days of receipt of a valid application. If the engineer defaults, or the purchaser interferes, the contractor may, without prejudice to any other remedies, either:

   (a) after giving the Purchaser or the Engineer 14 days' notice of his intention so to do, stop the Works or any part thereof until the said certificate be issued; in which case the expenses of the Contractor occasioned by such stoppage and the subsequent resumption of work shall be added to the Contract Price, or

   (b) after giving to the Purchaser or the Engineer one month's notice of his intentions so to do, terminate the Contract, whether or not the Contractor shall have stopped the Works in accordance with paragraph (a) or have given notice of his intention so to do.

- Clause 31(iv) – amounts in certificates, states:

   Every interim certificate shall certify:

   (a) the total amount to be paid as a stage payment in accordance with Clause 40 (Stage Payments) or the Contract Value of the Plant delivered to the Site, whichever is appropriate; and

   (b) the total value of the Works duly executed on the Site (excluding any amount included in (a) above);

   less the said total amount and value respectively so certified in the last previous certificate (if any), provided that the value of any Plant that, according to the decision of the Engineer, does not comply with the Contract shall not be included in any such certificate.

- Clause 31(v) – interim certificates not conclusive evidence of any matter.

Final certificates

Clauses 31(vi) to 31(ix) deal with final certificates. As with MF/1 these are not restricted to the whole of the works but apply also to (in G/90) 'portions'.

Clause 31(vi) states that final certificates can be applied for when the contractor ceases to have obligations for defects after taking-over. If renewals have proved necessary during the defects liability period final certificates can still be applied for except in respect of the renewed portions.

Clause 31(vii) requires the engineer to issue a final certificate within 28 days after receipt of a valid application. Unlike MF/1 there is no express statement to the effect that time only runs from receipt of all information reasonably required.

Clause 31(viii) requires a final certificate to certify the total of all amounts in interim certificates previously issued subject to authorized corrections or modifications.

Clause 31(ix) states that a final certificate is conclusive evidence of the value of the works unless proceedings are commenced by either party before the issue of the certificate or within one month thereafter. There is no express reference in G/90 to a final certificate being conclusive evidence of fulfilment of the contractor's performance.

Adjustments to certificates

Clause 31(x) requires amount falling due from the contractor to the purchaser to be deducted from the next certificate.

Clause 31(xi) permits the engineer to allow for corrections or modifications in previous certificates.

Payments due from the contractor

Clause 33 simply confirms the purchaser's right to make deductions from payments to the contractor in respect of amounts due from the contractor to the purchaser.

That is to say the purchaser is not bound to pay the amounts certified by the engineer in full if there is a contractual entitlement to a set-off.

Terms of payment (and retentions)

Clause 34(i) states times for payment and also indicates the percentages to be applied for retentions.

The Purchaser shall pay to the Contractor in the following manner the Contract Price adjusted to give effect to such additions thereto and such deductions therefrom as are provided for in these Conditions:

(a) Within 14 days from the presentation of each interim certificate a sum equal to 90 per cent of the sum certified therein.
(b) 95 per cent of the Contract Price adjusted as aforesaid within one month from the date certified in the taking-over certificate.
(c) The balance of the Contract Price adjusted as aforesaid within one month after the presentation of the final certificate provided that if the Contractor shall have furnished to the Purchaser a guarantee acceptable to the Purchaser for the repayment on demand of such balance he shall be entitled to payment thereof with or at any time after the payment provided for by paragraph (b) hereof.

Clause 34(ii) provides additional powers of retention in respect of specific defects.

If at any time at which any payment would fall to be made under paragraph (b) of Sub-Clause (i) of this clause there shall be any defect in any portion of the Works in respect of which such payment is proposed, the Purchaser may retain the whole of such payment provided that in the event of the said defect being of a minor character and not such as to affect the use of the Works, or the said portion thereof for the purpose intended without serious risk the Purchaser shall not retain a greater sum than represents the cost of making good the said minor defect. Any sum retained by the Purchaser pursuant to the provisions of this Sub-Clause shall be paid to the Contractor upon the said defect being made good.

Clause 34(iii) entitles the contractor to interest on late payments.

If the payment of any sum payable under Sub-Clause (i) of this clause shall be improperly delayed by the Purchaser or the Engineer interest on the amount of the delayed payment shall be added to the Contract Price for the period of the delay at the rate of 2 per cent per annum over the base rate, from time to time in force, of the Contractor's Bank named in the Appendix.

Clause 34(iv) provides the contractor with additional remedies for late payment.

If the Purchaser shall fail to make any payment as provided in this clause the Contractor shall have the like remedies, without prejudice to any other, as are provided in Sub-Clause (iii) of Clause 31 (Interim and Final Certificates).

The remedies in clause 31(iii) are:

- suspending progress on 14 days notice to the engineer;
- terminating the contract on one month's notice to the purchaser or the engineer.

Chapter 19

Defaults and termination

19.1 Introduction

This chapter deals with the circumstances in which contracts under the model forms can be prematurely terminated and with the defaults which justify the action of premature termination.

Broadly there are three sets of circumstances to consider:

- termination by agreement of the parties;
- termination arising from events beyond the control of the parties;
- termination resulting from the default of one of the parties.

It is also necessary to consider how the termination provisions of the model forms compare with termination rights at common law and the extent to which the model forms diminish those common law rights.

19.2 Termination/determination generally

Terminology

The model forms for plant contracts all refer throughout to 'termination of the contract'. Other standard forms use the phrases 'determination of the contract' or 'determination of the contractor's employment under the contract'.

Strictly speaking, the last phrase will generally be the correct description of what actually occurs when a contract is prematurely ended since the provisions of the contract usually remain in place to fix the consequences of the event. The word 'termination' is perhaps best used to describe ending by agreement; and 'determination' best used to describe ending as a result of one party's default.

Determination

The ordinary remedy for breach of contract is damages but there are circumstances in which the breach not only gives a right to damages but also entitles the innocent party to consider himself discharged from further performance.

Repudiation

Repudiation is an act or omission by one party which indicates that he does not intend to fulfil his obligations under the contract. In plant contracts, a contractor who abandons the works or a purchaser who refuses to give possession of the site are examples of a party in repudiation.

Determination at common law

When there has been repudiation or a serious breach which goes to the heart of the contract so that it is sometimes called 'fundamental' breach, common law allows the innocent party to accept the repudiation or the fundamental breach as grounds for determination of the contract. The innocent party would then normally sue for damages on the contract which had been determined.

The problems with common law determination are that it is valid only in extreme circumstances and it can readily be challenged.

Determination under contractual provisions

To extend and clarify the circumstances under which determination can validly be made and to regulate the procedures to be adopted, most standard forms of contracts include provisions for determination.

Many of the grounds for determination in standard forms are not effective for determination at common law. Thus failure by the contractor to proceed with due diligence and failure to remove defective work are often to be found in contracts as grounds for determination by the purchaser. But at common law neither of these will ordinarily be a breach of contract at all since the contractor's obligation is only to finish on time and to have the finished work in satisfactory condition by that time.

The commonest and the most widely used express provisions for determination relate to financial failures. Again at common law many of these are ineffective and even as express provisions they are often challenged as ineffective by legal successors of failed companies.

The very fact that grounds for determination under contractual provisions are wider than at common law leads to its own difficulties. A party is more likely to embark on a course of action when he sees his rights expressly stated than when he has to rely on common law rights. This itself can be an encouragement to error. Some of the best known legal cases on determination concern determinations made under express provisions but found on the facts to be lacking in validity.

In *Lubenham Fidelities* v. *South Pembrokeshire District Council* (1986) the contractor determined for alleged non-payment whilst the employer concurrently determined for failure to proceed regularly and diligently. On the facts, the contractor's determination was held to be invalid. But in *Hill & Sons Ltd* v. *London Borough of Camden* (1980), with a similar scenario, it was held on the facts that the contractor had validly determined.

Parallel rights of determination

Some contracts expressly state that their provisions, including those of determination, are without prejudice to any other rights the parties may possess. That is, the parties have parallel rights – those under the contract and those at common law – and they may elect to use either.

Other contracts are silent on the issue but the general rule is that common law rights can only be excluded by express terms. Contractual provisions, even though comprehensively drafted, do not imply exclusion of common law rights.

The point came up in the case of *Architectural Installation Services Ltd* v. *James Gibbons Windows Ltd* (1989) where it was held that while a notice of determination did not validly meet the timing requirements of the contractual provisions, nevertheless there had been a valid determination at common law. The general rule probably applies to MF/3 and G/90 but the exclusive remedies provisions of MF/1 and MF/2 (clauses 44.4 and 32.4 respectively) can be interpreted as exclusion of common law rights.

Frustration

At common law a contract is discharged and further performance excused if supervening events make the contract illegal or impossible or render its performance commercially sterile. Such discharge is known as frustration. A plea of frustration acts as a defence to a charge of breach of contract.

In order to be relied on, the events said to have caused frustration must be:

- unforeseen
- unprovided for in the contract
- outside the control of the parties
- beyond the fault of the party claiming frustration as a defence.

In the case of *Davis Contractors* v. *Fareham UDC* (1956) Lord Radcliffe said:

'Frustration occurs whenever the law recognises that without default of either party a contractual obligation has become incapable of being performed because the circumstances in which performance is called for would render it a thing radically different from that which was undertaken by the contract. *Non haec in foedera veni.* It was not this that I promised to do.'

In that case a contract to build 78 houses in eight months took 22 months to complete due to labour shortages. The contractor claimed the contract had been frustrated and he was entitled to be reimbursed on a *quantum meruit* basis for the cost incurred. The House of Lords held the contract had not been frustrated but was merely more onerous than had been expected.

The model forms for plant contracts do not contain 'frustration' provisions but MF/1 does have a *'force majeure'* clause which operates with similar effect.

Force majeure

The expression *'force majeure'* is of French origin. Under the French Civil Code *force majeure* is a defence to a claim for damages for breach of contract. It needs to be shown that the event:

- made performance impossible
- was unforeseeable
- was unavoidable in occurrence and effects.

In English law there is no doctrine of *force majeure*. Before 1863 and the case of *Taylor* v. *Caldwell* it was a rule of the law of contract that the parties were absolutely bound to perform any obligations they had undertaken and the fact that performance had become impossible did not provide relief from damages. In *Taylor* v. *Caldwell* a music hall which was to be hired for a concert was destroyed by fire the day before the performance; the court of Queen's Bench held the hirer not liable for damages by implying a term on impossibility of performance.

From this case developed the doctrine of frustration extending the sphere of impossibility to other instances of frustration. On basic legal principles, therefore, it is frustration and not *force majeure* which must be pleaded as a defence in English contract law.

Force majeure does, however, have a place in English law where it is expressly introduced as a contract term – as for example, in MF/1 where it provides grounds for extension of time and termination of the contract.

Force majeure excludes fault

Contractually based *force majeure* to be effective has to meet the same tests and has to conform with the doctrine of frustration in that there must be no fault attaching to the party using *force majeure* as a defence or a ground for claim. In *Sonat Offshore SA* v. *Amerada Hess Development Ltd* (1987) a *force majeure* clause entitled SONAT, an oil rig operation, to payment in certain circumstances. The clause applied '...when performance is hindered or prevented by strikes (except contractor induced strikes by contractor's personnel) or lockout, riot, war (declared or undeclared), act of God, insurrection, civil disturbances, fire, interference by any Government Authority or other cause beyond the reasonable control of such party...' Arising from the fault of SONAT there was an explosion and severe fire. The Court of Appeal held that '...other cause beyond the reasonable control...' did not include for negligence.

As a general rule, therefore, a party cannot rely on an event constituting *force majeure* within the meaning of such a clause unless it can be shown:

- the occurrence was beyond the control of the party claiming relief;
- there were no reasonable steps that party could have taken to avoid or mitigate the consequences of the event.

Additionally the courts are disposed to apply to *force majeure* clauses the same guidelines as they apply to the construction of clauses which purport to relieve a party from the consequences of his own negligence.

19.3 Defaults and termination in MF/1

Clause 49 of MF/1 deals generally with those contractor's defaults which entitle the purchaser to 'terminate the contract'.

Clause 50 of MF/1 deals with the contractor's bankruptcy and insolvency and clause 51 specifies those purchaser's defaults (including bankruptcy and insolvency) which entitle the contractor to 'terminate the contract'.

However, other clauses of MF/1 also refer to termination – although not all in the context of default. These clauses are:

- clause 8.2 – failure to provide bond or guarantee
- clause 25.5 – prolonged suspension
- clause 28.5 – consequences of failure to pass tests on completion
- clause 34.2 – prolonged delay
- clause 35.8 – consequences of failure to pass performance tests
- clause 40.3 – remedies on failure to certify or make payment
- clause 42.4 – infringement preventing performance
- clause 46.3 – termination for *force majeure*.

Clause 8.2 – failure to provide bond or guarantee

If the Contractor shall have failed to provide the bond or guarantee within 30 days after the date of the Letter of Acceptance or within such further period as may be advised by the Purchaser, the Purchaser shall be entitled to terminate the Contract by seven days notice to the Contractor. In the event of termination under this Clause the Contractor shall have no liability to the Purchaser other than to repay to the Purchaser all Costs properly incurred by the Purchaser incidental to the obtaining of new tenders.

Although failure to provide a bond or guarantee is a breach of contract which might well be sufficiently serious a breach to justify common law determination and full recovery of all the purchaser's damages this is not how it is regarded in MF/1.

The very restricted remedy available to the purchaser under clause 8.2 – costs incidental to obtaining new tenders – places this more in the category of termination by agreement.

Clause 25.5 – prolonged suspension

If suspension has continued for more than 90 days and the suspension is not necessitated by the reasons stated in Sub-Clause 25.4 (Disallowance of Additional Cost or Payment) the Contractor may by notice to the Engineer require him to give notice to proceed within 30 days.

If notice to proceed is not given within that time the Contractor may elect to treat the suspension as an omission under Clause 27 (Variations) of the part of the Works affected thereby. If the suspension affects the whole of the Works the Contractor may terminate the Contract in which event he shall be entitled to be paid in accordance with Sub-Clause 51.3 (Payment on Termination on Purchaser's Default) as if the Contract had been terminated under Sub-Clause 51.1 (Notice of Termination due to Purchaser's Default).

It is not wholly clear how this clause fits in with clause 46.3 (termination for *force majeure*).

Termination under clause 25.5 for prolonged suspension is treated as purchaser's default. But if the suspension arises from events detailed in clause 46.1 as *force majeure* it would appear that the termination should be treated on a no fault basis. To avoid the incompatibility engineers should be careful about instructing suspensions for any event which falls within the scope of clause 25.

Clause 28.5 – consequences of failure to pass tests on completion

If the Works or any Section fails to pass the Tests on Completion (including any repetition thereof) the Contractor shall take whatever steps may be necessary to enable the Works or the Section to pass the Tests on Completion and shall thereafter repeat them, unless any time limit specified in the Contract for the passing thereof shall have expired, in which case the Engineer shall be entitled to reject the Works or the Section and to proceed in accordance with Clause 49 (Contractor's Default).

This default is expressly listed in clause 49.1 as a default entitling the purchaser to terminate.

The wording of clause 28.5 is somewhat clumsy in implying that it is the engineer rather than the purchaser who terminates, but as mentioned in Chapter 9 this is likely to be corrected in the 1995 Revision 3 to MF/1.

One area of difficulty with clause 28.5 is that it treats the rejection of sections as though clause 49 deals with termination of sections. But clause 49 deals with termination of the contract not termination of sections of the contract. The result, which may be unintended, is that failure of a section entitles the purchaser to terminate the contract.

Clause 34.2 – prolonged delay

If any part of the Works in respect of which the Purchaser has become entitled to the maximum amount provided under Sub-Clause 34.1 (Delay in Completion) remains uncompleted the Purchaser may by notice to the Contractor require him to complete. Such notice shall fix a final Time for Completion which shall be reasonable having regard to such delay as has already occurred and to the extent of the work required for completion. If for any reason, other than one for which the Purchaser or some other contractor employed by him is responsible, the Contractor fails to complete within such time, the Purchaser may by further notice to the Contractor elect either:

(a) to require the Contractor to complete, or
(b) to terminate the Contract in respect of such part of the Works, and recover from the Contractor any loss suffered by the Purchaser by reason of the said failure up to an amount not exceeding the sum stated in the Appendix or, if no sum be stated that part of the Contract Price that is properly apportionable to such part of the Works as cannot by reason of the Contractor's failure be put to the use intended.

In this clause there is reference to termination of the contract in respect of parts – but no link is made with clause 49.

The damages recoverable by the purchaser for the default of prolonged delay are not restricted by the method of calculation used in clause 49 but they are limited in monetary value to stated sums or proportions of the contract price.

Clause 35.8 – consequences of failure to pass performance tests

If the Works or any Section fails to pass the Performance Tests (or any repetition thereof) within the period specified in the Special Conditions or, if no period is specified, within a reasonable time:

(a) - - - -
(b) - - - -
(c) where such failure of the Works or the Section would deprive the Purchaser of substantially the whole of the benefit thereof the Purchaser shall be entitled to reject the Works or the Section and to proceed in accordance with Clause 49 (Contractor's Default).

As with clause 25.8, this clause is expressly mentioned in clause 49.1. Again the problem of rejection of a section followed by termination of the contract exists in the wording of the two clauses.

Clause 40.3 – remedies on failure to certify or make payment

If the Engineer fails to issue a certificate of payment to which the Contractor is entitled or if the Purchaser fails to make any payment as provided in this Clause the Contractor shall be entitled:

(a) - - - –
(b) to terminate the Contract by giving 30 days notice to the Engineer and the Purchaser in any case where the Engineer has failed to issue a certificate of payment, whether or not the Contractor has previously stopped work under paragraph (a) of this Sub-Clause.

This is a provision entitling the contractor rather than the purchaser to terminate the contract.

As mentioned in Chapter 18 the contractor's right to terminate under this clause applies only when the engineer fails to issue a certificate – the contractor's right to terminate for non-payment on a certificate is found in clause 51.1(a) (notice of termination due to purchaser's default). But see the comment in that chapter on the likely amendment of this in the 1995 Revision 3 to MF/1.

As it currently stands the consequences in both cases are presumably intended to be the same – namely as set out in clause 51.3 (payment on termination due to purchaser's default). But clause 40.3(b) does not expressly refer to clause 51.3 nor does clause 51.1 include the clause 40(3)(b) default in its list of defaults.

Clause 42.2 – infringement preventing performance

If the Contractor shall be prevented from executing the Works, or the Purchaser is prevented from using the Works in consequence of any infringement of letters patent, registered design, copyright, trade mark or trade name and the party indemnifying the other in accordance with Sub-Clause 42.1 (Indemnity against infringement) is unable within 90 days after notice thereof from the other party to procure the removal at his own expense of the cause of prevention then:

(a) in the case of an infringement which is the subject of the Contractor's indemnity to the Purchaser under Sub-Clause 42.1 (Indemnity against Infringement) the Purchaser may treat such prevention as a default by the Contractor and exercise the powers and remedies available to him under Clause 49 (Contractor's Default), and

(b) in the case of an infringement which is the subject of the Purchaser's indemnity under Sub-Clause 42.3 (Purchaser's Indemnity against Infringement) the Contractor may treat such prevention as a default by the Purchaser and exercise the powers and remedies available to the Contractor under Clause 51 (Purchaser's Default).

These provisions are even handed between the parties allowing either to terminate in appropriate circumstances.

Clause 46.1 – *force majeure*

Force Majeure means:

– war, hostilities (whether war be declared or not), invasion, act of foreign enemies;

– ionizing radiations, or contamination by radio-activity from any nuclear fuel, or from any nuclear waste from the combustion of nuclear fuel, radio-active toxic explosive, or other hazardous properties of any explosive nuclear assembly or nuclear component thereof;

– pressure waves caused by aircraft or other aerial devices travelling at sonic or supersonic speeds;

– rebellion, revolution, insurrection, military or usurped power or civil war;

– riot, civil commotion or disorder;

– any circumstances beyond the reasonable control of either of the parties.

The only part of the clause which deserves comment is the event described as 'any circumstances beyond the reasonable control of either of the parties'.

As discussed in Chapter 11 a similar phrase (albeit in different circumstances) was given an unexpectedly wide meaning in the case of *Scott Lithgow Ltd* v. *Secretary of State for Defence* (1989).

In the context of this clause much depends upon how the word 'reasonable' is to be interpreted but there is a danger that either party could claim *force majeure* should they encounter difficulties beyond their direct control. A typical example might be insolvency of a sub-contractor. Could it be said that this was beyond the reasonable control of the contractor? In some cases – from a practical viewpoint – perhaps yes; in others, no. However, see the comments on the general legal interpretation of *force majeure* clauses in section 19.2 above.

Clause 46.2 – notice of *force majeure*

> If either party is prevented or delayed from or in performing any of his obligations under the Contract by *Force Majeure*, then he may notify the other party of the circumstances constituting the *Force Majeure* and of the obligations performance of which is thereby delayed or prevented, and the party giving the notice shall thereupon be excused the performance or punctual performance, as the case may be, of such obligations for so long as the circumstances of prevention or delay may continue.

As expressed here *force majeure* operates as a defence against non-performance.

But consider the application of this to non-payment by the purchaser on the grounds of *force majeure* (his banker perhaps having become insolvent). How then would the default provisions for non-payment apply? Clearly there is potential for conflict between the *force majeure* and the default provisions of MF/1 where the two overlap but different consequences are prescribed.

Clause 46.3 – termination for *force majeure*

> Notwithstanding that the Contractor may have been granted under Sub-Clause 33.1 (Extension of Time for Completion) an extension of the Time for Completion of the Works, if by virtue of Sub-Clause 46.2 (Notice of *Force Majeure*) either party shall be excused the performance of any obligation for a continuous period of 120 days, then either party may at any time thereafter, and provided such performance or punctual performance is still excused, by notice to the other terminate the Contract.

There are no requirements stated here for notice of intention to terminate and it is unlikely that the notice requirements of clauses 41 and 51 have any relevance since those clauses apply only to termination for default.

It would appear therefore that notice can be given promptly on the expiration of the 120 day period – which it appears from clause 46.2, runs from the date of giving notice of *force majeure*.

Clause 46.4 – payment on termination for *force majeure*

If the Contract is terminated under Sub-Clause 46.3 (Termination for *Force Majeure*) the Engineer shall certify, and the Purchaser shall pay to the Contractor in so far as the same shall not have already been included in certificates of payment paid by the Purchaser or be the subject of an advance payment, the Contract Value of the Works executed prior to the date of termination.

The Contractor shall also be entitled to have included in a certificate of payment and to be paid:

(a) the Cost of materials or goods reasonably ordered for the Works or for use in connection with the Works which have been delivered to the Contractor or of which the Contractor is legally liable to accept delivery. Such materials or goods shall become the property of the Purchaser when paid for by the Purchaser. The Purchaser shall be entitled to withhold payment in respect thereof until such goods or materials have been delivered to or to the order of, the Purchaser;

(b) the amount of any other expenditure which in the circumstances was reasonably incurred by the Contractor in the expectation of completing the whole of the Works;

(c) the reasonable Cost of removal of Contractor's Equipment and the return thereof to the Contractor's works in his country or to any other destination at no greater Cost;

(d) the reasonable Cost of repatriation of all the Contractor's staff and workmen employed at the Site on or in connection with the Works at the date of such termination.

The opening lines of this clause have to be taken with some flexibility since strictly if the contract has been terminated there is no remaining power for the engineer to certify.

But that apart, the scheme for payment is clearly that the contractor should recover only his costs and that the purchaser has no liability for such other damages as would be due for breach of contract.

Clause 49.1 – contractor's default

If the Contractor shall assign the Contract, or sub-let the whole of the Works without the consent of the Purchaser, or if the Engineer has rejected the Works or a Section under Sub-Clauses 28.5 (Consequences of Failure to Pass Tests on Completion) or 35.8 (Consequences of Failure to Pass Performance Tests) or shall certify that the Contractor:

(a) has abandoned the Contract, or

(b) has without reasonable excuse suspended the progress of the Works for 30 days after receiving from the Engineer written notice to proceed, or

(c) despite previous warnings in writing from the Engineer is not executing the

Works in accordance with the Contract, or is failing to proceed with the Works with due diligence or is neglecting to carry out his obligations under the Contract so as to affect adversely the carrying out of the Works, then the Purchaser may give 21 days' notice to the Contractor of his intention to proceed in accordance with the provisions of this Clause. Upon the expiry of such notice the Purchaser may without prejudice to any other remedy under the Contract forthwith terminate the Contract and enter the Site and expel the Contractor therefrom but without thereby releasing the Contractor from any of his obligations or liabilities which have accrued under the Contract and without affecting the rights and powers conferred by the Contract on the Purchaser or the Engineer. Upon such termination the Purchaser may himself complete the Works or may employ any other contractor so to do, and the Purchaser shall have the free use of any Contractor's Equipment for the time being on the Site.

The list of defaults given in clause 49.1 is similar to that found in other plant, process and construction contracts. However the purchaser would be unwise to assume that in acting on any of the specified defaults he would automatically be within his rights under the contract.

Termination, or determination of the contractor's employment under the contract which is what clause 49.1 describes, is a matter so serious that it will be scrutinized by the courts if challenged to ensure that the provisions have been followed to the letter and that the default alleged can be properly substantiated.

For generalized defaults such as abandonment and failing to proceed with due diligence, proper substantiation can be a difficult problem.

'Abandoned the Contract'

Abandonment may be patently obvious where the contractor has left the site or has otherwise given notice of his intention not to fulfil his obligations. Such abandonment is repudiation and grounds for common law determination.

However, there can also be situations where the contractor's conduct in ceasing work is neither abandonment nor repudiation. In *Hill* v. *Camden* (1981), the contractor, unable to obtain prompt payment cut his staff and labour. The employer took this action as repudiatory and served notice of determination. Lord Justice Lawton had this to say of the contractor's action:

'The plaintiffs did not abandon the site at all; they maintained on it their supervisory staff and they did nothing to encourage the nominated subcontractors to leave. They also maintained the arrangements which they had previously made for the provision of canteen facilities and proper insurance cover for those working on the site.'

Failing to proceed with due diligence

This is a matter which should be approached with the greatest of caution. The courts have been most reluctant to impose on contractors any greater obligation than to finish on time.

In *Greater London Council* v. *Cleveland Bridge* (1986) the Court of Appeal refused to imply a term into a building contract that the contractor should proceed with due diligence notwithstanding the inclusion of that phrase in the determination clause of that contract. The point was repeatedly made in that case that the contractor should be free to programme his work as he thought fit. For other cases showing the difficulty of defining due diligence and failure to proceed with it, see *Hill* v. *Camden* (1981) and *Hounslow* v. *Twickenham Garden Developments* (1971).

Failure by the contractor to proceed in accordance with his approved clause 14 programme might provide some evidence to support a charge of failing to proceed with due diligence although failure to proceed to the programme is not itself a breach of contract.

The reference to 'previous warnings' in clause 49(1) should, perhaps, be read in conjunction with clause 46. That clause places a duty on the engineer to notify the contractor if he considers progress too slow to achieve completion by the due date. However see the case of *Tara Civil Engineering Ltd* v. *Moorfield Developments Ltd* (1989) where the court refused to go behind the certificate of an engineer setting in motion the determination provisions of an ICE contract to see if other provisions of the contract had been applied.

Assignment and sub-letting

For comment on assignment see Chapter 7.

As to sub-letting note that clause 49.1 lists sub-letting the whole of the works without the consent of the purchaser as a specified default. Under clause 3.2 (sub-contracting) the contractor is to obtain the consent of the engineer to sub-letting of any part of the works but breach of this is not a specified default.

Clause 49.2 – valuation at date of termination

As soon as practicable after the Purchaser has terminated the Contract the Engineer shall, by or after reference to the parties and after making such enquiries as he thinks fit, value the Works and all sums then due to the Contractor as at the date of termination in accordance with the principles of Clause 39 (Certificates and Payment) and certify the amount thereof. The amount so certified is herein called 'the Termination Value'.

It can be argued that to determine the 'Termination Value' in accordance with the principles of clause 39 is inappropriate since valuation under clause 39 is based on the assumption that the works will be completed by the contractor.

However, any over assessment of the termination value is accounted for in clause 39.3 within the cost of completion. But, in any event, it is understood that the 1995 Revision 3 to MF/1 will clarify the position by making it clear that the engineer is to value only the part of the works executed prior to termination.

Clause 49.3 – payment after termination

> The Purchaser shall not be liable to make any further payments to the Contractor until the Costs of execution and all other expenses incurred by the Purchaser in completing the Works have been ascertained and the amount payable certified by the Engineer (herein called 'the Cost of Completion'). If the Cost of Completion when added to the total amounts already paid to the Contractor as at the date of termination exceeds the total amount which the Engineer certifies would have been payable to the Contractor for the execution of the Works, the Engineer shall certify such excess and the Contractor shall upon demand pay to the Purchaser the amount of such excess. Any such excess shall be deemed a debt due by the contractor to the Purchaser and shall be recoverable accordingly. If there is no such excess the Contractor shall be entitled to be paid the difference (if any) between the Termination Value and the total of all payments received by the Contractor as at the date of termination.

The two important aspects of this clause are:

- the purchaser is not required to make any payments after termination until completion has been achieved and its costs are known;
- the contractor is liable to the purchaser for the costs of completing.

The questions which are not answered within the clause are:

- does the contractor have any other liability to the purchaser for damages for breach of contract?
- does the limitation of the contractor's liability as stated in the appendix apply to termination?

The answers are to be found in clauses 44.3 and 44.4. It would appear that the contractor's liability is limited to the costs of completion and to the value stated in the appendix (or the contract price if not stated).

Clause 50 – bankruptcy and insolvency

If the Contractor shall become bankrupt or insolvent, or have a receiving order made against him or compound with his creditors, or being a corporation commence to be wound up, not being a members' voluntary winding up for the purpose of amalgamation or reconstruction, or have an administration order made against him or carry on his business under an administrator or a receiver or manager for the benefit of his creditors or any of them, the Purchaser shall be entitled:

(a) to terminate the Contract forthwith by notice to the Contractor or to the administrator, receiver, manager or liquidator or to any person in whom the Contract may become vested, in which event the provisions of Clause 49 (Contractor's Default) shall apply, or
(b) to give such administrator, receiver, manager or liquidator or other person the option of carrying out the Contract subject to his providing a guarantee for the due and faithful performance of the Contract up to an amount to be agreed.

This provision wisely gives the purchaser a choice of action on the contractor's insolvency.

Other standard forms which provide for automatic termination on insolvency have run into various legal difficulties – not least whether performance bonds can be called in such circumstances – see the case of *Perar BV* v. *General Surety and Guarantee Co Ltd* (1994).

Clause 51.1 – notice in the event of purchaser's default

In the event of the Purchaser:

(a) failing to pay to the Contractor the amount due under any certificate of the Engineer within 30 days after the date of its issue subject to any deduction that the Purchaser is entitled to make under the Contract, or
(b) interfering with or obstructing the issue of any certificate of the Engineer, or
(c) becoming bankrupt or (being a corporation) going into liquidation other than for the purpose of a scheme of reconstruction or amalgamation, or carrying on its business under an administrator, receiver, manager or liquidator for the benefit of its creditors or any of them, or
(d) appointing a person to act with or in replacement of the Engineer against the reasonable objections of the Contractor,

the Contractor shall be entitled without prejudice to any other rights or remedies under the Contract [and in respect of paragraph (a) above in addition to the provisions of Sub-Clause 40.3 (Remedies on Failure to Certify or Make Payment)] to terminate the Contract by giving 14 days' notice to the Purchaser with a copy to the Engineer.

Not all model forms expressly state the purchaser's defaults which provide the contractor with entitlement to terminate.

Of those that do, items (a), (b) and (c) above are usually included. Item (d) however is unusual. Since it is not clear what is meant by the phrase 'to act with ... the Engineer' (see Chapter 8) any contractor should proceed with the greatest caution in this matter.

Clause 51.2 – removal of contractor's equipment

> Upon the giving of notice under Sub-Clause 51.1 (Notice of Termination due to Purchaser's Default) the Contractor shall with all reasonable despatch remove from the Site all Contractor's Equipment.

This provision which places the contractor under an obligation to remove his equipment from site upon notice of termination may go some way to resolving a difficult question which arises frequently in termination clauses. Whether the giving of notice of termination is provisional on the default being remedied or whether the notice is absolute and the time for remedying is passed.

This wording of clause 51.2 suggests that notice is absolute in respect of the purchaser's defaults but it does not automatically follow that the same applies to the contractor's defaults.

Clause 51.3 – payment on termination due to purchaser's default

> In the event of termination under Sub-Clause 51.1 (Notice of Termination due to Purchaser's Default) the Engineer shall act as provided in Sub-Clause 49.2 (Valuation at Date of Termination) and certify the Termination Value of the Works as at the date of termination. The Engineer shall, on the application of the Contractor accompanied by supporting details, also certify the amount of any expenditure reasonably incurred by the Contractor in the expectation of the performance of, or in consequence of the termination of, the Contract to the extent that the same has not been included in the Termination Value. The Engineer shall also certify in respect of the Contractor's loss of anticipated profit on the Contract the percentage referred to in Sub-Clause 41.2 (Allowance for Profit on Claims) on the difference between the total of the Termination Value plus the expenditure before referred to and the Contract Price but in no case shall the total amounts so certified exceed the Contract Price. Thereafter the Engineer shall issue a certificate of payment for the amount by which the said Termination Value, expenditure and allowance for profit exceeds the total of sums previously paid to the Contractor and such certificate of payment shall be paid by the Purchaser within 30 days after the date of issue.

The intention of this clause is that the contractor should recover for the purchaser's default:

- the value of work undertaken prior to termination, plus
- the costs incurred by the contractor in expectation of completion, plus
- loss of profit on work not undertaken.

There is, however, a potential difficulty in that the total amount recoverable is limited to the contract price – which by definition is the tender price. Clearly this may give inadequate recompense to the contractor if variations have been ordered prior to termination.

19.4 Defaults and termination in MF/2

The principal differences between MF/2 and MF/1 on the subjects of defaults and termination arise because:

- MF/2 has no provisions for *force majeure*
- MF/2 has no provisions for performance tests.

Other differences are that the following clauses are so worded that they provide the purchaser with remedies for the contractor's defaults which fall short of termination:

- clause 17.5 – failure of tests or inspections
- clause 24.2 – prolonged delay.

The clauses of MF/2 which do expressly refer to termination are:

- clause 8.2 – failure to provide bond or guarantee
- clause 19.5 – prolonged suspension
- clause 28.3 – failure to certify or make payment
- clause 30.4 – infringement preventing performance.

The clauses for contractor's default (clauses 34 and 35) and purchaser's default (clause 36) are virtually the same as in MF/1 with delivery substituted for completion.

19.5 Defaults and termination in MF/3

MF/3 does not refer to defaults and termination as such. Instead it refers to rejection and replacement.
   The relevant clauses are:

- clause 4.3 – failure on testing
- clause 5.1 – notice of rejection

- clause 5.2   – consequences of rejection
- clause 7.4   – prolonged delay.

Clause 4.3 – failure on testing

If on a test made pursuant to Sub-Clause 4.2 (Prescribed Tests) the goods or any part thereof shall fail to pass the prescribed tests or to give the specified performance such goods or part thereof shall, if the Vendor so desires, be tested again or the Vendor may submit for test other goods in their place. If the goods or the said other goods shall fail to pass the test or to give the specified performance, the Purchaser shall be entitled by notice in writing to reject the goods or such part thereof as shall have failed as aforesaid.

Clause 5.1 – notice of rejection

The Purchaser shall be entitled, by notice in writing given within a reasonable time after delivery, to reject goods delivered which are not in accordance with the Contract.

The important questions here are:

- What is a reasonable time after delivery?
- Does the phrase 'goods . . . not in accordance with the Contract' apply to defects of no real consequence?

The answer to the first question is that what is reasonable is a matter of fact to be determined on the circumstances of each case. The answer to the second question is that it is probably subject to the above test of reasonableness.

Clause 5.2 – consequences of rejection

When goods have been rejected, either under Clause 4 (Tests) or Sub-Clause 5.1 (Notice of Rejection), the Purchaser shall be entitled, provided he does so without undue delay, to replace the goods so rejected. There shall be deducted from the Contract Price that part thereof which is properly apportionable to the goods rejected. The Vendor shall pay to the Purchaser any sum by which the expenditure reasonably incurred by the Purchaser in replacing the rejected goods exceeds the sum deducted. All goods obtained by the Purchaser to replace rejected goods shall comply with the Contract and shall be obtained at reasonable prices and, when reasonably practicable, under competitive conditions. Where goods have been rejected as aforesaid the Vendor shall not be under any liability to the Purchaser except as provided in this clause and as may arise under Clause 7 (Time for Delivery).

This provision, although dealing with the vendor's default, does not as might be expected operate to the purchaser's advantage. Instead it acts as a limitation of liability for the vendor and prevents the purchaser from recovering the full damages for breach of contract to which he might otherwise be entitled.

Clause 7.4 – prolonged delay

This clause is quoted in Chapter 11.

As with clause 5.2 it limits the vendor's liability for his default to replacement costs.

19.6 Defaults and termination in G/90

The clauses of G/90 expressly covering defaults and/or termination are:

- clause 3(ii) – failure to provide security for due performance
- clause 12 – contractor's default
- clause 13 – bankruptcy
- clause 31(iii)(b) – interference with or failure to issue interim certificates
- clause 34(iv) – failure to make payment.

Other clauses which touch upon the issues without expressly mentioning termination are:

- clause 14(vi) – rejection of plant
- clause 15(iv) – delayed plant
- clause 29(ii) – prolonged suspension.

Clause 3(ii) – failure to provide security for due performance

If the guarantee or other security for the due performance of the Contract required to be furnished pursuant to this clause shall not be duly furnished by the Contractor to the Purchaser within one month after the Contract has been entered into, the Purchaser may, at his option, without prejudice to any rights or claims he may have against the Contractor by reason of the Contractor's non-compliance with any of the provisions of this clause, and within seven days after the expiry of the said period, by notice in writing to the Contractor terminate the Contract forthwith, and the Purchaser shall thereupon not be liable for any claim or demand from the Contractor in respect of anything then already done or furnished, or in respect of any other matter or thing whatsoever, in connection with the Contract, but the Purchaser shall be entitled to be repaid by the Contractor all out-of-pocket expenses properly incurred by the Purchaser incidental to the obtaining of new tenders.

This clause is not restrictive of the purchaser's rights as is clause 8.1 of MF/1. In G/90 the costs of obtaining new tenders are clearly in addition to, and not the limitation of, the purchaser's right to damages for breach of contract.

Clause 12 – contractor's default

If the Contractor shall neglect to execute the Works with due diligence and expedition, or shall refuse or neglect to comply with any reasonable orders given him in writing by the Engineer in connection with the Works, or shall contravene the provisions of the Contract, the Purchaser may give seven days' notice in writing to the Contractor to make good the failure, neglect, or contravention complained of. Should the Contractor fail to comply with the notice within seven days from the date of service thereof in the case of a failure, neglect, or contravention capable of being made good within that time or otherwise within such time as may be reasonably necessary for making it good, then and in such case the Purchaser shall be at liberty to employ other workmen, and forthwith execute such part of the Works as the Contractor may have neglected to do, or, if the Purchaser shall think fit, it shall be lawful for him, without prejudice to any other rights or remedies the Purchaser may have, to take the Works wholly or in part out of the Contractor's hands and either by his own workmen or by re-contracting with any other person or persons to complete the Works or any part thereof, and in either event the Purchaser shall have the free use of all Contractor's Equipment that may be on the Site in connection with the Works, without being responsible to the Contractor for fair wear and tear thereof, and to the exclusion of any right of the Contractor over the same, and the Purchaser shall be entitled to retain and apply any balance which may be otherwise due on the Contract by him to the Contractor, or such part thereof as may be necessary to the payment of the cost of executing the said part of the Works or of completing the Works as the case may be. If the cost of completing the Works or executing a part thereof as aforesaid shall exceed the amount that would otherwise become due to the Contractor in accordance with the Contract, the Contractor shall pay such excess. If the Purchaser pursuant to this clause takes the Works or part thereof out of the Contractor's hands, the Contractor's liability under Clause 26 (Delay in Completion) shall immediately cease in respect of the Works or part thereof, without prejudice to any such liability that shall have already accrued.

This is an unusually worded default clause which owes its origins to the old MF 'A'. Although it does not refer expressly to termination or expulsion of the contractor from the site it has much to recommend it in clarity in the way it expresses its intentions on what should be done.

This can be summarized as:

- contractor's default
  ↓
- purchaser gives seven days notice to the contractor to make good
  ↓
- contractor fails to comply
  ↓
- purchaser at liberty to take work out of contractor's hands
  ↓
- purchaser entitled to free use of contractor's equipment on the site
  ↓
- purchaser entitled to retain any payments due to the contractor until the costs of completion are known
  ↓
- purchaser to pay any balance due to the contractor
  ↓
- contractor not liable for damages for late completion after work taken out of his hands.

'Without prejudice to any other rights and remedies'

G/90 does not have the same exclusive remedy provisions as MF/1 and one interpretation of the phrase in clause 12 'without prejudice to any other rights and remedies the Purchaser may have' is that the purchaser is entitled to recover damages for breach of contract over and above those of the costs of completion. However, these would probably be subject to the limitations in clause 22 on loss of use, loss of profit and loss of other contracts.

Clause 13 – bankruptcy

If the Contractor shall become bankrupt or insolvent, or have a receiving order made against him, or compound with his creditors, or being a corporation commence to be wound up, not being a members' voluntary winding up for the purpose of reconstruction or amalgamation, or carry on its business under a receiver for the benefit of its creditors or any of them, the Purchaser shall be at liberty either:

(a) to terminate the Contract forthwith by notice in writing to the Contractor or to the receiver or liquidator or to any person in whom the Contract may become vested, and to act in the manner provided in Clause 12 (Contractor's Default) as though the last mentioned notice had been the notice referred to in such clause and the Works had been taken out of the Contractor's hands, or

(b) to give such receiver, liquidator, or other person the option of carrying out the Contract subject to his providing a guarantee for the due and faithful performance of the Contract up to an amount to be agreed.

Clause 31(iii)(b) – interference with or failure to issue interim certificates

If the Engineer shall fail to issue an interim certificate as provided in this clause or if the Purchaser shall interfere with or obstruct the issue of any of such certificate the Contractor may, without prejudice to any other remedy, either:

(a) after giving the Purchaser or the Engineer 14 days' notice of his intention so to do, stop the Works or any part thereof until the said certificate be issued; in which case the expenses of the Contractor occasioned by such stoppage and the subsequent resumption of work shall be added to the Contract Price, or
(b) after giving to the Purchaser or the Engineer one month's notice of his intentions so to do, terminate the Contract, whether or not the Contractor shall have stopped the Works in accordance with paragraph (a) hereof or have given notice of his intention so to do.

There is no indication in clause 31(iii)(b) of the amount of damages recoverable by the contractor for the purchaser's default as found in MF/1.

A point of interest here is that in G/90 the limitations on liability set out in clause 22 apply only to the contractor's liability whereas in MF/1 the similar provisions in clause 44.2 limit the liability of both the contractor and the purchaser.

Clause 34(iv) – failure to make payment

If the Purchaser shall fail to make any payment as provided in this clause the Contractor shall have the like remedies, without prejudice to any other, as are provided in Sub-Clause (iii) of Clause 31 (Interim and Final Certificates).

The remedies here are either to stop work or terminate the contract.

Chapter 20

Settlement of disputes

20.1 Introduction

Because only the parties to a contract can sue on the contract a dispute in the formal sense is only a dispute between the parties. Consequently formal dispute resolution procedures such as arbitration and litigation involve only the contractor and the purchaser.

But, in practice, of course, contractors under plant, process and construction forms frequently find themselves in dispute with the engineer or his representative in matters such as workmanship, testing, valuations and certificates. Purchasers also can find themselves in dispute with their own appointed engineers when they consider that the engineer, in exercising his discretion, has been over generous to the contractor. There is also the question of disputes between main contractors and sub-contractors.

Aspects of disputes

Contractual provisions for the settlement of disputes therefore need to consider, if they are to be comprehensive:

- how disputes between the contractor and the engineer's representative are to be resolved;
- how disputes between the contractor and the engineer are to be resolved;
- the extent to which unresolved disputes between the contractor and the engineer should be treated as disputes between the contractor and the purchaser;
- whether the purchaser should have the same rights as the contractor in disputing decisions and certificates of the engineer;
- how disputes between the contractor and the purchaser should be resolved;
- the extent to which disputes between contractors and sub-contractors should be joined with disputes between contractors and purchasers.

Meaning of disputes

Most standard forms, including the model forms for plant contracts, deal reasonably clearly with the detail of dispute resolution procedures; usually by way of referrals for reconsideration for disputes with the engineer and arbitration for disputes between the parties. But the wider issues of how and when disputes with the engineer become disputes between the parties and how extensive are the rights of the purchaser to instigate disputes are frequently poorly addressed.

The reason for this latter inadequacy may be that there is some underlying conflict between the two issues involved. If the engineer is seen solely as the agent of the purchaser then clearly any disputes between the contractor and the engineer amount to disputes between the contractor and the purchaser; and there is no place in the contract for disputes between the purchaser and the engineer. But if the engineer is seen as truly independent, then it is less clear why disputes between the contractor and the engineer should be regarded as disputes between the contractor and the purchaser; but there is a better case for the purchaser to have the same rights as the contractor.

Contracts which state that an arbitrator has power to open up and review any decisions, certificates, etc of the engineer avoid some of the above difficulties. But otherwise since the engineer is usually required to combine the functions of agent of the purchaser and independent contract administrator there is ample scope for disputes on the meaning of disputes unless the contract deals in detail with the finality of and/or challenges to the engineer's exercise of his authority.

The model forms for plant contracts are not without their difficulties in these matters.

20.2 Dispute resolution generally

Arbitration is the traditional method of resolving disputes in commercial contracts and most standard forms contain arbitration provisions.

In recent years however various criticisms of arbitration have gained ground, in particular, that:

- it is too adversarial;
- it is too costly and time consuming;
- it is not appropriate for disputes requiring a prompt decision.

Alternative dispute resolution

Such views, which are by no means confined to the United Kingdom, have led to the exploration of new ideas and the development of new techniques for dispute resolution. They include mini trials, conciliation and mediation and collectively they are known as alternative dispute resolution (ADR).

The mini trial

The mini trial, or executive tribunal as it is sometimes called, is a non-binding procedure at which the parties present their case before senior executives from both parties assisted by a neutral expert. The role of the expert who may have a legal or technical background is to help the parties focus on critical issues and if appropriate to offer advice on the merits or likely outcome of the dispute if a settlement is not reached.

Conciliation and mediation

The procedure whereby a neutral person assists the parties to reach a settlement to their dispute can be known as either conciliation or mediation. The terms are not fixed and there is a good argument that flexibility should be maintained in the use of the terms so that the various procedures they encompass can converge and adapt to suit particular circumstances.

Nevertheless, there is a move, at least in this country, to ascribe the terms conciliation and mediation to distinct procedures. A mediator is expected to take a more active role than a conciliator.

Dispute resolution panels

These are panels set up at the commencement of a contract to rule on all disputes arising. Usually the panel members are independent of the parties but this is not essential.

Amicable settlement

All disputes are open to amicable settlement but some contracts formalize the process and describe how it should be conducted.

Adjudication

Adjudication now features in many standard forms of contract. The adjudicator may be appointed on an *ad hoc* basis as and when disputes arise or appointed at the outset depending on the terms of the contract. Frequently the role of the adjudicator is confined to valuations and/or to dispute resolution during the progress of the works.

Reference to experts

The IChemE model forms have an unusual method for resolving certain specified disputes. They are referred to an independent expert whose decision is final and binding and cannot be challenged in any subsequent proceedings.

Reference to the engineer

Some contracts require disputes to be referred to the engineer to the contract for a decision before any proceedings can be commenced.

Objectives of dispute resolution procedures

Not all the methods of dispute resolution listed above have the same objectives. Broadly they can be divided into categories for:

- resolving disputes until completion is reached;
- conditions precedent to arbitration;
- providing alternatives to arbitration.

Finality

Arbitration and litigation produce legally binding and enforceable decisions. The problem with other methods of dispute resolution is that they rarely do so. This means that for finality to be achieved by alternative dispute resolution methods the parties must generally be satisfied with the outcome.

20.3 Arbitration generally

Arbitrations in England and Wales are regulated by the Arbitration Act 1950 as amended by the Arbitration Act 1979. Section 32 of the 1950 Act defines an 'arbitration agreement' as a 'written agreement to submit present or future differences to arbitration, whether an arbitrator is named therein or not'.

Enforcing the arbitration agreement

If there is a valid arbitration agreement covering the dispute the courts will normally stay any legal proceedings brought by one party, thus upholding the agreement – section 4 Arbitration Act 1950. The courts will not grant a stay however if the applicant for the stay has taken a step in the legal action, that is, shown himself willing to contest the legal proceedings.

In order to obtain a stay the application must:

- show there is an arbitration clause in the contract;
- make an application at an early stage in the proceedings;
- establish that he is ready and willing to arbitrate.

Of course the parties may, if they are of the same mind, avoid the arbitration agreement altogether and go to court.

Multi-party actions

In some disputes there may be numerous parties involved – contractor, purchaser, sub-contractors, suppliers, designers. But arbitration, unlike litigation, cannot cope with multi-party disputes unless all the parties to the proceedings so agree or unless the arbitration agreement requires the use of rules of procedure which cater for multi-party actions.

Since different sets of proceedings on the same dispute would involve extra cost, procedural difficulties and a grave risk of inconsistency the courts may exercise their discretion and refuse a stay.

The model form of sub-contract for use with MF/1 states, as do various other forms of sub-contract, that the contractor is entitled to require sub-contract disputes to be joined with any associated main contract arbitration.

However, these clauses usually rely on the purchaser's agreement to the arbitration becoming multi-party and only rarely, and certainly not in MF/1, is there any express obligation in the main contract that the purchaser must allow sub-contractors to be joined in if the contractor so requires.

Scope of the arbitration

The arbitrator's jurisdiction derives from the arbitration agreement and the courts distinguish between disputes arising 'under' a contract from disputes arising 'in connection with' a contract. Arbitration agreements in modern contracts are generally drafted with wide wording to ensure the agreement is not limited in its scope.

The question of whether an arbitrator can consider matters which strike at the very existence of the contract – misrepresentation and the like – has been in debate for many years. The point being that if the arbitrator rules there is no contract he has apparently lost the authority on which to make his ruling.

However, in *Ashville Investments* v. *Elmer Contractors* (1987) the Court of Appeal held that claims arising out of alleged innocent or negligent misstatements were claims arising 'in connection with' the contract and the arbitrator had jurisdiction.

Control by the courts

The courts have a general supervisory role over arbitrations and under powers given by the 1950 Act the courts may:

- appoint an arbitrator if the parties cannot agree;
- make orders for security and examination on oath;
- remove an arbitrator who is guilty of delay, misconduct or bias;
- remit an award to an arbitrator for his reconsideration;
- set aside an award for misconduct.

Under the 1979 Act the courts may:

- review arbitration awards;
- determine preliminary points of law;
- extend the time for commencement;
- enforce awards.

Appeals to the courts

Section 1(2) of the 1979 Act permits appeals on questions of law arising out of an award with the consent of the parties or with the leave of the court. Leave is not granted unless the court considers that the determination of the question concerned could substantially affect the rights of one of the parties. Generally leave to appeal will only be granted where there is a standard form of contract involved. The point in issue must be one of public interest. In one-off contracts leave to appeal will only be granted in special circumstances.

Under the 1979 Act the parties can enter into an exclusion agreement which precludes rights of appeal on points of law.

Conduct of the arbitrator

An arbitrator is bound by two principal rules:

- to act fairly between the parties in accordance with the rules of natural justice;
- to decide only the matters submitted to him in accordance with the arbitration agreement.

Failure to act fairly can lead to removal of the arbitrator by the courts for what is termed 'misconduct'. Examples include:

- an arbitrator making an interim award on a point of law without hearing one of the parties – *Modern Engineering (Bristol)* v. *Miskin & Co* (1981);
- an arbitrator failing to give a party an opportunity to deal with his (the arbitrator's) own special knowledge of the facts – *Fox* v. *P. G. Wellfair Ltd* (1981);
- an arbitrator receiving a prejudicial document from one party and refusing to let the other party have a copy of it – *Maltin Engineering* v. *Dunne Holdings* (1980).

If the arbitrator exceeds his jurisdiction the courts will set aside his award as in *Secretary of State for Transport* v. *Birse-Farr Joint Venture* (1993).

Characteristics of arbitration

Attempts are frequently made to list the advantages and disadvantages of arbitration compared with litigation. The exercise is largely pointless because individual experiences will always suggest different conclusions.

However, arbitration does have characteristics which distinguish it from litigation, some of which may be of advantage; some of disadvantage. It depends upon what the parties are seeking.

The principal characteristics of arbitration are as follows:

- the hearing is private;
- the award is published privately;
- the parties may be able to choose their own arbitrator;
- the venue of the hearing is flexible;
- the hours, date and all administrative details of the hearing are flexible;
- the hearing can be less formal than in litigation;
- the rights of appeal against an award are restricted;
- not ideal for multi-party disputes;
- the arbitrator and the venue have to be paid for;
- lawyers are optional;
- lay advocates are allowed;
- the arbitrator has no general power to order security for costs but may have under some rules;

- a reluctant party may be less robustly dealt with;
- can be cheaper and quicker;
- documents-only may be a permitted procedure;
- the arbitrator has power to award interest and costs;
- the arbitrator can make site visits, etc;
- the arbitrator may be given powers the courts do not have;
- arbitration may be imposed if an agreement is included in the contract;
- the legal aid scheme is not applicable.

Arbitration rules

The more commonly used building and civil engineering forms of contract, JCT and ICE, refer in their arbitration clauses to particular sets of rules for the conduct of arbitrations – JCT Rules and ICE Rules respectively. Both include short procedures designed to keep costs and time to the minimum. In the absence of prescribed rules the provisions of the Arbitration Acts 1950–1979 apply.

The model forms for plant contracts do not refer to any particular rules in their standard text but the special conditions could state which rules should apply.

Commencement of proceedings

Arbitration proceedings commence when one party serves on the other a notice requiring a dispute to be referred to arbitration. In general, there is no set form for such notices but standard arbitration rules do sometimes detail a procedure which includes serving:

- notice of dispute
- notice to refer to arbitration
- notice to concur in the appointment of an arbitrator.

The advantage of formalizing matters in this way is that the arbitrator's terms of reference are likely to be better defined if proceedings are preceded by a notice of dispute than if a simple notice to refer is the starting point.

20.4 Settlement of disputes in MF/1 and MF/2

Clause 52 (disputes and arbitration) sets out the principal provisions in MF/1 for the settlement of disputes but it is also necessary to consider:

- clause 2.6 – disputing engineer's decisions, instructions and orders, and
- clause 39.12 – effect of final certificate of payment.

MF/2 has similar provisions to MF/1 in clauses 37, 2.6 and 27.12.

Clause 2.6 of MF/1

This clause has been considered in detail in Chapter 8. It provides for the contractor to dispute certain decisions, instructions and orders of the engineer by referring the matter in question to the engineer. Unless either party challenges the decision given by the engineer on such referral within a prescribed timescale, the decision of the engineer is final and binding on the parties.

Clause 39.12 of MF/2

This clause has been considered in detail in Chapter 18. It provides that a final certificate of payment shall be conclusive evidence of the value of the works and that the contractor has performed his obligations unless proceedings are commenced within a prescribed time.

Clause 52.1 of MF/1 – notice of arbitration

If at any time any question, dispute or difference shall arise between the Purchaser and the Contractor in relation to the Contract or in any way connected with the Works, which cannot be settled amicably, either party shall as soon as reasonably practicable give to the other notice of the existence of such question, dispute or difference specifying its nature and the point at issue, and the same shall be referred to the arbitration of a person to be agreed upon. Failing agreement upon such person within 30 days after the date of such notice, the arbitration shall be conducted by some person appointed on the application of either party by the President of the Institution named in the Appendix (or by his deputy appointed by such President for the purpose). A question, dispute or difference relating to a decision, instruction or order of the Engineer shall not be referred to arbitration except in accordance with Sub-Clause 2.6 (Disputing Engineer's Decisions, Instructions or Orders).

'If at any time'

These words indicate that arbitration proceedings can be commenced prior to completion of the works and at any time afterwards within the statutory limitation periods.

However the effects of the final certificate and other clauses of MF/1 relating to defects and exclusive remedies are such that the full benefits of statutory limitation periods (six years for contracts under hand; twelve years for contracts executed as deeds) are rarely available to either party.

'Any question, dispute or difference'

The words 'question' and 'difference' are probably superfluous.

There is, in any event, some doubt on whether a question can properly be referred to arbitration. Essentially arbitration deals with the situation when one party makes an assertion which the other party rejects.

'In relation to the Contract'

A more common phrase in arbitration clauses is 'in connection with or arising out of the contract'.

'In any way connected with the Works'

This phrase, it is suggested, cannot be taken literally.

There might well be a dispute between the parties connected with the works which has nothing to do with the particular MF/1 contract.

'Which cannot be settled amicably'

These words create unnecessary uncertainty in the clause. They do not set out any defined procedure for amicable settlement and it is not clear whether they are intended as a firm condition precedent to arbitration. But they do give a party the opportunity to argue that the notice to refer is premature in that amicable settlement is still possible thereby creating an effective delaying tactic.

'As soon as reasonably practicable'

It is not clear what the consequences of a party failing to give notice 'as soon as reasonably practicable' would be. Two possibilities exist, either the dispute can no longer be pursued, or the arbitration clause itself is no longer effective.

Since an arbitration clause can be seen as an exclusion of the right to litigate, which must therefore be strictly interpreted, it is unlikely that such a vague phrase would be effective in shutting a party out of any legal remedy.

'Specifying its nature and the point at issue'

A recipient of a notice which fails to comply with the above is entitled to reject it. For example, a vague notice of the type 'We require all disputes on this contract to be referred to arbitration' would clearly be a defective notice.

'Failing agreement ... within 30 days'

This is quite a short period for the parties to agree upon an arbitrator after service of a notice to refer.

The provision does not apply strictly in the sense that if the parties fail to agree within the 30 days the arbitrator 'shall' be an appointed person. What it intends is that if the parties do not agree within 30 days either party is entitled (but not obliged) to apply to the president of the nominated body for an appointment.

'A decision, instruction or order of the Engineer'

These words apply to matters which are required to be dealt with in accordance with clause 2.6.

In so far that they relate to the carrying out of the works their intention is fairly clear in both clauses 2.6 and 52.1. Namely, that if the contractor wishes to dispute any such matter he must do so within a set timescale.

The problem is that whilst the context of clause 2.6 is the carrying out of the works the words themselves might be capable of wider application. For example, is it to be taken that a 'decision, instruction or order of the Engineer' under clause 2.6 is to include the issue of, and matters relating to, certificates or the granting of extensions of time and the like.

There are two arguments against such a proposition. Firstly, that clause 2.6 allows only the contractor to make the initial challenge – which suggests that it is limited to matters of practical effect. And secondly, the wider interpretation taken in conjunction with rules in clause 52.1 would act as an exclusion of the purchaser's rights which is anything but clearly expressed.

Scope of clause 52.1

If the words 'decision, instruction or order of the Engineer' exclude matters relating to certificates, extensions of time and the like there is no problem with clause 52.1. A dispute on such matters is not then caught by the reference to clause 2.6 and the dispute can clearly be referred to arbitration by either party.

However if the words do include matters relating to certificates, extension of time, etc then disputes on such matters are outside the scope of clause 52.1 unless they have first been disputed by the contractor under clause 2.6. This has two possible effects:

- either certificates, extensions of time, etc granted by the engineer are final and binding on the parties unless challenged by the contractor, or
- disputes on such matters cannot be raised by the purchaser through the provisions of the contract and they are outside the scope of the arbitration agreement – which leaves open the question of whether the purchaser has recourse to litigation.

Clause 52.2 – performance to continue during arbitration

> Performance of the Contract shall continue during arbitration proceedings unless the Engineer shall order the suspension thereof. If such suspension be ordered the additional Costs to the Contractor occasioned by such suspension shall be added to the Contract Price. No payment due or payable by the Purchaser shall be withheld on account of a pending reference to arbitration.

Unlike some standard forms MF/1 does not require arbitration proceedings to be deferred until completion of the works.

Such a requirement would not be appropriate because MF/1 does not have any procedures similar to adjudication to deal with matters which require decisions to progress the works.

'The additional Costs ... shall be added to the Contract Price'

These words are not qualified by the usual proviso of the type 'unless any such suspension is necessary due to default of the Contractor'. It would therefore appear that even if the contractor loses his case in arbitration he is entitled to the costs of any suspension arising from his raising the dispute.

'No payment due or payable ... shall be withheld'

These words apply to a 'pending reference to arbitration' – whatever that might be. The intention may be to restrict the purchaser's rights of set-off in respect of disputed sums.

20.5 Settlement of disputes in MF/3

Clause 18.1 provides a simple arbitration agreement.

Clause 18.1 – arbitration

> If at any time any question, dispute, or difference whatsoever shall arise between the Vendor and the Purchaser upon, in relation to, or in connection with the Contract, either of them shall give to the other notice in writing of the existence of such question, dispute, or difference, and the same shall be referred to the arbitration of a person to be agreed upon or failing agreement within 14 days after the date of such notice, of some person to be appointed, on the application of either party, by the President for the time being of the Institution named in the Appendix (or by his deputy appointed by such President for the purpose).

Note that there is no timing requirement between the notice of dispute and the notice to refer unless the clause is to be interpreted so that the notice of dispute is the notice to refer.

That may be the correct interpretation having regard to the phrase 'such notice'. But if it is correct the time period of 14 days from giving notice of dispute to concurring in the appointment of an arbitrator is obviously inadequate.

20.6 Settlement of disputes in G/90

The arbitration provisions of G/90 are very close in wording to those of MF/1.

Clause 37 – arbitration

> 37(i) If at any time any question, dispute or difference shall arise between the Purchaser and the Contractor, either party shall, as soon as reasonably practicable, give to the other notice in writing of the existence of such question, dispute or difference specifying its nature and the point at issue, and the same shall be referred to the arbitration of a person to be agreed upon, or failing such agreement within six weeks, to some person appointed on the application of either of the parties hereto by the President for the time being of the Institution named in the Appendix, provided that a question, dispute or difference relating to a decision, instruction, or order of the Engineer shall not be referred to arbitration unless notice has been given by the Contractor in accordance with Clause 19(b) (Engineer's Decisions).

37(ii) Performance of the Contract shall continue during arbitration proceedings unless the Engineer shall order the suspension thereof or of any part thereof, and if any such suspension shall be ordered the reasonable expenses of the Contractor occasioned by such suspension shall be added to the Contract Price. No payments due or payable by the Purchaser shall be withheld on account of pending reference to arbitration.

Points to note

- There is no reference to disputes and the like arising out of or being in connection with the contract.
- The time for concurring in the appointment of an arbitrator is six weeks.
- Only the contractor's notice under clause 19(b) is mentioned.
- Only the reasonable expenses of the contractor are recoverable in the event of suspension.

Clause 19(b) restrictions

Under clause 19(b) either party may refer a disputed decision, instruction or order of the engineer to arbitration. However, under clause 37(1) arbitration is expressly precluded if the contractor fails to give notice in accordance with clause 19(b).

As with MF/1 this raises important questions as to the scope of the words 'decision, instruction or order of the Engineer' – found in G/90 in both clauses 19(b) and 37(i) – and as to the exclusion of the purchaser's rights to challenge any such decision, instruction or order.

See the comment in section 20.4 on these points but note that in G/90 clause 19 commences with the words 'The Contractor shall proceed with the Works in accordance with decisions, instructions and orders given by the Engineer'. This opening, it is suggested, adds heavily to the argument that the clause is concerned only with decisions, instructions and orders of practical effect and that it does not extend to matters related to certificates, extensions of time and the like.

Table of cases

Note: end references are to chapter sections.
The following abbreviations of law reports are used:

AC	Law Reports Appeal Cases Series
ALJR	Australian Law Journal Reports
All ER	All England Law Reports
BLR	Building Law Reports
CILL	Construction Industry Law Letter
CLD	Construction Law Digest
Const LJ	Construction Law Journal
Con LR	Construction Law Reports
C & P	Carrington & Payne's Reports
DLR	Dominion Law Reports
Exch	Exchequer Reports
Giff	Gifford's Reports
HBC	Hudson's Building Contracts
KB	Law Reports, King's Bench Division
LGR	Local Government Reports
LJQB	Law Journal, Queen's Bench
Lloyd's Rep	Lloyd's List Law Reports
NZLR	New Zealand Law Reports
SALR	South African Law Reports
STARK	Starkie's Reports
TLR	Times Law Reports
TR	Term Reports
WLR	Weekly Law Reports
WR	Weekly Reporter

Table of clause references

Index

Note: references are to chapter sections.

411